Environmental Sustainability Issues in the South Texas–Mexico Border Region

David Ramirez • Jianhong Ren
Kim D. Jones • Harriet Lamm
Editors

Environmental Sustainability Issues in the South Texas–Mexico Border Region

Springer

Editors
David Ramirez
Department of Environmental Engineering
Texas A&M University-Kingsville
Kingsville, Texas 78363, USA

Kim D. Jones
Department of Environmental Engineering
Texas A&M University-Kingsville
Kingsville, Texas 78363, USA

Jianhong Ren
Department of Environmental Engineering
Texas A&M University-Kingsville
Kingsville, Texas 78363, USA

Harriet Lamm
Texas A&M Engineering Experiment Station
Partner Outreach
College Station, Texas 77845-3470, USA

Logos reprinted with permission

ISBN 978-94-007-7121-5 ISBN 978-94-007-7122-2 (eBook)
DOI 10.1007/978-94-007-7122-2
Springer Dordrecht Heidelberg New York London

Library of Congress Control Number: 2013947815

Printed on acid-free paper

Springer is part of Springer Science+Business Media (www.springer.com)

Acknowledgements

This material is based upon work supported by the National Science Foundation under Grant No. HRD-0734850. Any opinions, findings, and conclusions or recommendations expressed in this material are those of the authors and do not necessarily reflect the views of the National Science Foundation.

Special thanks to the Center for Research Excellence in Science and Technology (CREST) – Research on Environmental Sustainability of Semi-Arid Coastal Areas (RESSACA) funded by the National Science Foundation, and the Institute for Sustainable Energy and the Environment at Texas A&M University-Kingsville.

For more information contact:
CREST – RESSACA
Texas A&M University-Kingsville
700 University Blvd., MSC-213
Kingsville, TX 78363
Phone: 361-593-5556
Fax: 361-593-2069

Contents

Contributors

Marcia Baeza Department of Environmental Engineering, Frank H. Dotterweich College of Engineering, Texas A&M University-Kingsville, Kingsville, TX, USA

Wayne Belzer Environmental Management Division, International Boundary and Water Commission, United States Section, Clean Rivers Program, El Paso, TX, USA

Jhumoor Biswas Indian Institute of Social Welfare and Business Management (IISWBM), Kolkata, West Bengal, India

Lee W. Clapp Department of Environmental Engineering, Frank H. Dotterweich College of Engineering, Texas A&M University-Kingsville, Kingsville, TX, USA

Brian Dyson Department of Environmental Engineering, Frank H. Dotterweich College of Engineering, Texas A&M University-Kingsville, Kingsville, TX, USA

U.S. Environmental Protection Agency, Washington, DC, USA

Zuber Farooqui Department of Mechanical and Energy Engineering, University of North Texas, Denton, TX, USA

California Air Resources Board, Sacramento, CA, USA

Brianna Flores Department of Environmental Engineering, Frank H. Dotterweich College of Engineering, Texas A&M University-Kingsville, Kingsville, TX, USA

Surface Mining and Reclamation Division, Railroad Commission of Texas, Austin, TX, USA

Yaneth P. Gamboa Department of Environmental Engineering, Frank H. Dotterweich College of Engineering, Texas A&M University-Kingsville, Kingsville, TX, USA

Prajay Gor Department of Environmental Engineering, Frank H. Dotterweich College of Engineering, Texas A&M University-Kingsville, Kingsville, TX, USA

Laura Guerrero-Medrano Instituto Tecnológico y de Estudios Superiores de Monterrey (ITESM), Monterrey, Nuevo León, Mexico

School of Medicine and Health Sciences, Tecnologico de Monterrey, Monterrey, Nuevo León, Mexico

Kuruvilla John Department of Mechanical and Energy Engineering, University of North Texas, Denton, TX, USA

Kim D. Jones Department of Environmental Engineering, Frank H. Dotterweich College of Engineering, Texas A&M University-Kingsville, Kingsville, TX, USA

Institute for Sustainable Energy and the Environment, Texas A&M University-Kingsville, Kingsville, TX, USA

Sushma Krishnamurthy School of Sciences, University of Louisiana at Monroe, Monroe, LA, USA

Harriet Lamm TAMUS LSAMP Project, Texas A&M Engineering Experiment Station, College Station, TX, USA

Alvaro Martinez Department of Environmental Engineering, Frank H. Dotterweich College of Engineering, Texas A&M University-Kingsville, Kingsville, TX, USA

Gerardo M. Mejia-Velazquez Instituto Tecnológico y de Estudios Superiores de Monterrey (ITESM), Monterrey, Nuevo León, Mexico

Center for Environmental Quality, School of Engineering and Information Technologies, Tecnologico de Monterrey, Monterrey, Nuevo León, Mexico

Celia de la Mora Orozco Institute for Sustainable Energy and the Environment, Texas A&M University-Kingsville, Kingsville, TX, USA

National Research Institute for Forestry, Agriculture and Animal Production, Mexico D.F., Mexico

Department of Environmental Engineering, Frank H. Dotterweich College of Engineering, Texas A&M University-Kingsville, Kingsville, TX, USA

Jayanth Kumar Pudota Department of Environmental Engineering, Frank H. Dotterweich College of Engineering, Texas A&M University-Kingsville, Kingsville, TX, USA

David Ramirez Department of Environmental Engineering, Frank H. Dotterweich College of Engineering, Texas A&M University-Kingsville, Kingsville, TX, USA

Center for Research Excellence in Science and Technology – Research on Environmental Sustainability of Semi-Arid Coastal Areas (CREST-RESSACA), Texas A&M University-Kingsville, Kingsville, TX, USA

Jianhong Ren Department of Environmental Engineering, Frank H. Dotterweich College of Engineering, Texas A&M University-Kingsville, Kingsville, TX, USA

Ruth Reyna-Caamaño Instituto Tecnológico y de Estudios Superiores de Monterrey (ITESM), Monterrey, Nuevo León, Mexico

Center for Environmental Quality, School of Engineering and Information Technologies, Tecnologico de Monterrey, Monterrey, Nuevo León, Mexico

Amrutha Sakaray Department of Environmental Engineering, Frank H. Dotterweich College of Engineering, Texas A&M University-Kingsville, Kingsville, TX, USA

FutureNet Group, Inc, Detroit, MI, USA

Jesus Santos-Guzman Instituto Tecnológico y de Estudios Superiores de Monterrey (ITESM), Monterrey, Nuevo León, Mexico

School of Medicine and Health Sciences, Tecnologico de Monterrey, Monterrey, Nuevo León, Mexico

Neelesh Sule Providence Engineering & Environmental Group, Irving, TX, USA

Thomas C. Vaughan Department of Biology and Chemistry, Texas A&M International University, Laredo, TX, USA

Yungang Wang Environmental Energy Technologies Division, Lawrence Berkeley National Laboratory, Berkeley, CA, USA

Yifang Zhu Environmental Health Sciences Department, Jonathan and Karin School of Public School, University of California, Los Angeles, Los Angeles, CA, USA

Center for Clean Air in the Institute of Environment and Sustainability, University of California, Los Angeles, Los Angeles, CA, USA

Chapter 1
Introduction to Environmental Sustainability Issues in the South Texas–Mexico Border Region

Harriet Lamm, David Ramirez, and Jianhong Ren

Abstract Due to the importance of the Rio Grande in providing water to this semi-arid ecoregion and the major farming areas and cities along its banks, studies are being done to help assure the safety of the water flowing in the river. Studies are also being done along the coastal regions to monitor the lagoons and inland waterways connected with the Gulf of Mexico. Water pollution can come from runoff from irrigation and rain as well as from discharge from the factories located along the waterways, feeder creeks, and rivers. Air pollution comes from the factories on both sides of the Rio Grande, along the coast where refineries and other industries are located near the ports, and from dirt being carried by strong winds from farmland and the sand dune areas. Dust also comes from other parts of the world as has been seen with dust storms originating in the Sahara that propel the dust into the jet stream which then drops the dust over this region. The chapters within this book focus on air and water pollution along the Rio Grande, its watershed, and the coastal region due to the fragile ecology of the region.

H. Lamm (✉)
TAMUS LSAMP Project, Texas A&M Engineering Experiment Station,
College Station, TX 77845-3470, USA
e-mail: hlamm@tamu.edu

D. Ramirez
Department of Environmental Engineering, Frank H. Dotterweich College of Engineering,
Texas A&M University-Kingsville, Kingsville, TX 78363, USA

Center for Research Excellence in Science and Technology – Research on Environmental
Sustainability of Semi-Arid Coastal Areas (CREST-RESSACA), Texas A&M
University-Kingsville, Kingsville, TX 78363, USA

J. Ren
Department of Environmental Engineering, Frank H. Dotterweich College of Engineering,
Texas A&M University-Kingsville, Kingsville, TX 78363, USA

D. Ramirez et al. (eds.), *Environmental Sustainability Issues in the South Texas–Mexico Border Region*, DOI 10.1007/978-94-007-7122-2_1,

Keywords South Texas • Northeastern Mexico • Semi-arid • Ecoregions • Rio Grande Plains • South Texas Plains • Coastal Plains • Tamaulipan mezquital • Coahuila • Nuevo Leon • Tamaulipas • Rio Grande • Gulf Coast • Laredo • Nuevo Laredo • Brownsville • Matamoros

1.1 Area

The South Texas–Mexico border region includes South Texas and northeastern Mexico. The area begins about latitude 29°30′N to 25°30′N and longitudes 101°30′ to 99°30′ and 97°15′ up the Gulf coast to 96°30′ (Rand McNally & Company, 2000) or an area of about 141,500 km^2 (54,600 mi^2). The western and northwestern boundary in both Mexico and South Texas is the southern and eastern edge of the Chihuahuan desert. The northern boundary in Texas is the Balcones escarpment (Texas hill country) and is roughly the area just south of line from Del Rio to San Antonio. The eastern boundary is the oak prairies lying along a line from to San Antonio to Victoria to the coast at Port Lavaca and Port O'Conner and then southward along the Gulf of Mexico and the coastal prairies, tidal areas, and barrier islands. The southern and western boundary in Mexico is the Sierra Madre Oriental range (Cook, Adams, Valero, Schipper, & Allnutt, 2001; Hoekstra et al., 2010).

South Texas is also referred to as the Rio Grande Plains, South Texas Plains, and Coastal Plains. Depending on the source of information, the area includes from 35 to 39 counties and this varies even more when talking to residents of the area (South Texas, n.d.; "South Texas Plains Wildlife District". Texas Parks and Wildlife Department [TPWD], 2008b). Figure 1.1 gives the perspective of the area of the state considered to be South Texas with respect to the entire state.

1.2 Cities and Populations

In the South Texas area along the Rio Grande, cities and industry are sparse to non-existent until reaching the area just west of Reynosa and Roma (Table 1.1). From this point until the river reaches the Gulf, the population has increased until it appears to be one city the entire distance. Both industry and populations have raised the amount of pollution that becomes air borne and contaminates run-off from rain. Surface water is affected as a result of waste water treatment plants along this part of the river as some of the treated water is released into the river and closer to the coast into the coastal marshes and lagoons (Rio Grande Valley, n.d.; South Texas, n.d.).

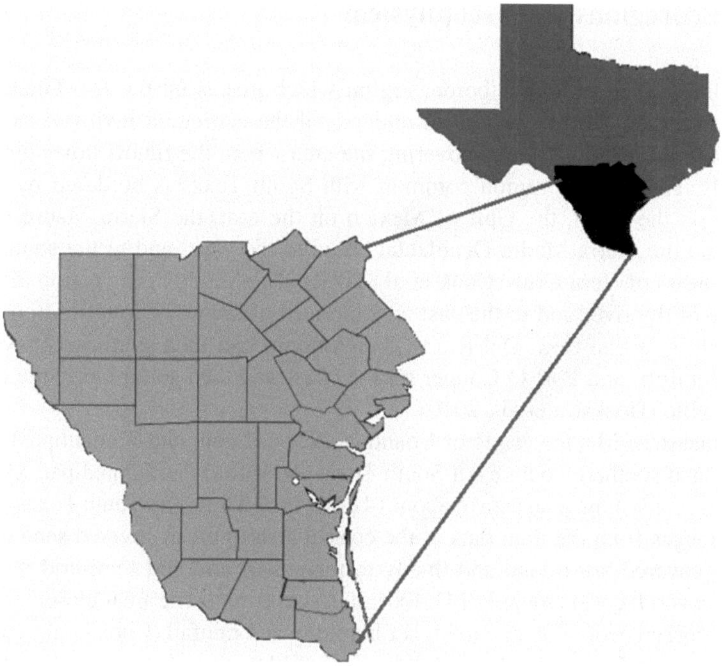

Fig. 1.1 Counties within area considered to be South Texas

Table 1.1 Cities of South Texas (United States Census Bureau, 2010)

South Texas city	2010 census population	Size rank	County
San Antonio	1,359,758	2	Bexar
Corpus Christi	307,953	8	Nueces
Laredo	241,935	10	Webb
Brownsville	178,430	16	Cameron
McAllen	133,742	20	Hidalgo
Mission	79,368	43	Hidalgo
Edinburg	79,147	44	Hidalgo
Pharr	72,513	47	Hidalgo
Harlingen	66,122	51	Cameron
Victoria	63,131	54	Victoria
Weslaco	36,721	88	Hidalgo
San Juan	34,872	91	Hidalgo
Schertz	32,478	97	Bexar
Eagle Pass	26,807	109	Maverick
Kingsville	26,322	110	Kleberg
Seguin	25,943	114	Guadalupe
San Benito	24,729	116	Cameron

1.3 Ecoregions and Geophysical

When looking at the larger border region which makes up the Rio Grande river basin watershed, it is important to understand the ecoregion involved as well as the ecoregion of South Texas covering the area where the runoff flows directly to the Gulf. The Mexico region common with South Texas is bordered by the Rio Grande on the north, the Gulf of Mexico on the east, the Sierra Madre Oriental range and the Sierra Madre Occidental range to the west, and to the south by the moist forests of Vera Cruz (Cook et al., 2001). The South Texas region along this expanse of the river and to the east is comprised of about 39 counties bounded on the north by a line from Del Rio to San Antonio and then southeast to Victoria, Point Comfort, and Port O'Conner on the coast and then south along the coast to Brownsville (Hoekstra et al., 2010).

Northeastern Mexico (parts of Coahuila, Nuevo Leon, and Tamaulipas) and the western and southern counties of South Texas are within the Tamaulipan Mezquital ecoregion consisting of an area of about 141,500 km^2. The entire South Texas–Mexico region ranges from the tidal flats of the coastal area to brush covered sand dunes to densely covered brush land and finally to regions of arid and semi-arid plants and ground cover. It covers roughly 141,400 km^2 (54,600 mi^2). The areas closer to the Gulf are in a coastal ecoregion due to higher humidity and rainfall (Cook et al., 2001).

The state of Tamaulipas begins at the Gulf of Mexico and moves to the northwest to just north of Nuevo Laredo and Laredo. The state of Nuevo Leon runs only a short distance along the Rio Grande north of Nuevo Laredo and Laredo before the state of Coahuila begins. Coahuila is bounded by the Rio Grande on the north from here to a point south of the Chisos Mountains in the Big Bend that is about one-half the river distance between Boquillas (Boquillas del Carmen) and Castolon. From this point west and northwest is the state of Chihuahua until reaching Cuidad Juarez, El Paso, and New Mexico (Rand McNally & Company, 2000).

The Chihuahuan Desert covers most of Chihuahua and the extreme western edge of Coahuila. It crosses the Rio Grande into Texas west of the Pecos River that enters the Rio Grande north of Del Rio ("Chihuahuan Desert," World Wildlife Organization, 2001). This is farther north and west of what is considered South Texas (Cook et al., 2001).

The ecoregion of this semi-desert area is home to plants and animals not found in other areas due to the climate and rainfall. The Tamaulipan Mezquital (also referred to as the Coahuila Desert) and the Deserts of Nuevo Leon influence those Mexican states as well as the western counties of South Texas (roughly west of U.S. Highway 281 and south of U.S. Highway 90) (Olson et al., 2001). A map depicting this area (referred to as the South Texas Plains Ecoregion) can be viewed on the Texas Parks and Wildlife Department web site at http://www.tpwd.state.tx.us/publications/pwdpubs/pwd_pl_w7000_1187a/media/4.pdf. It continues its eastward movement each year as indicated by plant life found only in arid and semi-arid regions beginning to appear in these areas. The extreme and exceptional drought which has been occurring on both sides of the border has allowed the desert to move more quickly—in feet rather than inches per year as indicated by the lack of vegetation in pastures and the surviving plants (Cook et al., 2001).

The Rio Grande begins its southward journey in Colorado and travels through New Mexico before reaching Mexico and Texas. The amount of water making it past El Paso is dependent on winter snowfall, the water usage for drinking and agriculture, and the amounts held in man-made lakes along its course (Mertz, 2012). From El Paso to Del Rio, the river supplies water to agriculture on both sides as well as industries located across from Presidio in Ojinaga. Another possible source of river pollution may come from the La Jitas and Terlingua areas where mercury mines were located many years ago and where the raw and cooked ore sits on the surface.

Along the Texas-Mexico region of the Rio Grande, there are 3 major rivers in the 12 rivers which feed into the Rio Grande from Mexico and 2 from Texas. Some of the rivers on the Mexico side have dams on them to help control water flow into the Rio Grande. The Rio Grande has two dams on this specific stretch of the river–Amistad Dam (just north of Del Rio and Acuna) and Falcon Dam (just west of Zapata). The states of northeastern Mexico which supply and take water from the Rio Grande are Coahuila (northern most), Tamaulipas, and Nuevo Leon on the Gulf coast. Their major cities along the river are Cuidad Acuna, Piedras Negras, Nuevo Laredo, Reynosa, and Matamoras. The corresponding Texas cities are Del Rio, Eagle Pass, Laredo, McAllen, and Brownsville with many other cities between McAllen and Brownsville that also influence the river's flow. Texas' only two major rivers feeding into the Rio Grande are the Pecos River and the Devil's River, both north of Del Rio. There are many creeks and one short, spring-fed river which feed into the Rio Grande (Rand McNally & Company, 2000). The two dams control the water so that there is a constant flow downstream to the Gulf of Mexico. Demands from agriculture and municipal use decrease the flow of the river as it flows southward so that the flow into the Gulf may almost cease which was the case during 2001–2003. Due to excessive drought and water usage a sand bar formed at the mouth of the river so that no flow entered the Gulf until the river flooded again. The sand bar caused by exceptional drought and water usage was removed by the Army Corp of Engineers in 2003 so flow into the Gulf could resume (Rio Grande, n.d.).

The Gulf coast has beaches on the barrier islands of Matagorda, San Jose, Mustang, and Padre with an inter-coastal waterway running between the barrier islands and the mainland. There are cuts in the barrier islands where natural passages do not exist to allow ships access to the Gulf from the ports at Matagorda, Port Aransas, Port Mansfield, and Port Isabel. Bays allow natural access to the port in Corpus Christi and Port O'Conner. The Port of Corpus Christi has a constant flow of tankers bringing petroleum and other goods and materials from other countries to the refineries and docks located on the port's channel (South Texas, n.d.).

Moving north and east from the Rio Grande into South Texas, one finds a region bounded to the northwest and north by the Texas hill country (Balcones escarpment) and on the northeast by the oak prairies. The northern boundary lies along a line from Del Rio to San Antonio to Victoria to the Gulf Coast at Port Lavaca and Port O'Conner. Leaving Port O'Conner one would journey south-southwest along the coast to Port Isabel. The land along this eastern side is mostly coastal plains ranging from coastal marsh flats to sand dunes to low-lying, flat, oak mott savannahs to fertile farm land and then back to coastal marshes and sand dunes farther south. As one moves west from the coast, the amount of yearly rainfall decreases as the distance increases. The eco-region also changes as one travels westward from Port O'Conner and southwestward toward the Rio Grande. Upon crossing each of the rivers flowing into the Gulf from this region, changes in vegetation and geological terrain are evident. The most noticeable change occurs west of the Nueces River (the start of the Tamaulipan Mezquital) in the area referred to as the Nueces Strip which encompasses land from the Nueces to the Rio Grande. This part of the region has a colorful history that matches the changes in the terrain from coastal prairies with fertile farmland to semi-arid brush lands to the sand dunes of the Wild Horse Desert and to the fertile flood plains of the Rio Grande in the Lower Rio Grande Valley (referred to as the Valley) (South Texas, n.d.; United States Census Bureau, 2010).

The farther west you travel, the more rugged the terrain becomes as it becomes more barren and desert-like, especially west from Uvalde, Cotulla, Freer, and Hebbronville. Desert terrain, topology, and plants begin about 20 miles west of Uvalde marking the edge of the Tamaulipan Mezquital (Spearing, 1991). This is the area most affected by the encroachment of the Tamaulipan Mezquital or the Coahuila Desert. During wet years, the area is lush with grass while during most years it is covered with mesquite, huisache, prickly pear, and other plants common to semi-arid and desert-like regions (Cook et al., 2001). Along the coast the humidity is constant across the coastal plains which range from fertile farm soil to vegetation-covered sand dunes ("Habitats," TPWD, 2008a).

The South Texas–Mexico border region's industries include ranching, farming, petro-chemical plants, and the mining of several different minerals. Lignite coal, uranium, caliche, Fuller's earth, asphalt, limestone, sulfur, salt, and clay are among the list of naturally found minerals that are mined. Some semi-precious stones (e.g., quartz and common opal), sand, gravel, and sea shell are also produce of this region (Garner, 2012). In the Mexican state of Coahuila, the major mineral products include coal, iron, titanium, feldspar, lead, and dolomite. Agricultural products from these arid and semi-arid regions include cotton, wines, brandies, and timber. Producing 95 % of Mexico's coal makes it the home of the automobile industry factories, specifically the plants for General Motors and Chrysler (Coahuila, n.d.). The region also produces various vegetables and fruits sold throughout the country (Cook et al., 2001).

The petroleum industry currently is focused on the region just east of the Rio Grande from Eagle Pass to Laredo and eastward to Three Rivers, Kenedy, Nixon, Cuero, Gonzales, and points farther northeast which lay over the Hawkville (Eagle Ford Shale) formation (Eagle Ford Shale Trend Map, 2011). This area's productivity is a result of the underlying geological formations originating from the Mesozoic (upper Cretaceous) to the Cenozoic (lower and upper Tertiary and Pleistocene) layers of the earth. The land formation running south from Laredo consists of Eocene age tan sandstones. The terrain in the southern portion of this region (Lower Rio Grande Valley) belongs to the Holocene (Cenozoic) age

(Spearing, 1991). Oil and gas exploration have occurred and are still occurring on both sides of the Rio Grande but currently are being increased on the Texas side due to the unstable circumstances existing along the U.S.-Mexico border. The increase in and intensity of the drilling in the Hawkville formation has raised concerns on water quality and quantity as well as the effects on the Carrizo-Wilcox and Gulf Coast aquifers. Concerns center on the water resources necessary for the fracturing process that "brings in" a well and the effects of the process on the surrounding geological structures. Recycling of this water has been discussed. The *San Antonio Express News* reported in summer 2011 that two portable "processing" plants were being brought to the Kenedy area to clean up the frac water that is returned to the surface. The recycled water from the two mobile water purification centers will result in fresh water and clean concentrated salt water. This will allow the water to be reused in the process and decrease the depletion of the aquifers (Hamilton, 2011, July 3).

1.4 Summary

Due to the importance of the Rio Grande in providing water to this semi-arid ecoregion with its major farming areas and cities along the river, studies are being done to help assure the safety of the water flowing in the river. Studies are also being done along the coastal regions to monitor the lagoons and inland waterways connected with the Gulf of Mexico. Water pollution can come from runoff from irrigation and rain as well as from discharge from the factories located along the waterways, feeder creeks, and rivers. Air pollution comes from the factories on both sides of the Rio Grande, along the coast where refineries and other industries are located near the ports, and from dirt being carried by strong winds from farmland and the sand dune areas. Dust also comes from other parts of the world as has been seen with dust storms originating in the Sahara that propel the dust into the jet stream which then drops the dust over this region. The chapters within this book focus on the effects of air and water pollution along the Rio Grande, its watershed, and the coastal region to the fragile ecology of the this South Texas-Mexico region.

Air quality questions are also being raised due to the increases in traffic, drilling, small separating plants located on the increasing number of pipelines, and well flares in areas that formerly had low traffic flows and less gas and petroleum production. Air quality is as important as water quality and has controllable and uncontrollable pollutants. Industries and municipal traffic flows are major sources of air pollution in this region. Agriculture also contributes to the fine particulate matter pollution as well as the thousands of miles of dirt and caliche roads traveled in the region. Drought conditions have increased this particular pollutant significantly due to the loss of the vegetative groundcover and lack of soil moisture.

The research studies on air quality in the South Texas – Mexico border region where conducted at locations along the Rio Grande, an urban coastal municipality, and four select interior cities. Dr. Alvaro Martinez and his graduate students did two studies on the effects of pollution to a person's health. The first study focused on the prevalence of asthma, allergic rhinitis, and eczema in school children in cities of the Northeastern Mexican State of Tamaulipas to determine if there were potential correlations with ambient pollution factors. The second study centered along the U.S.-Mexico border region that has experienced high economic and population growth assessed the impacts of air quality on public health for this region by means of environmental health indicators.

Dr. Yifang Zhu led an intensive continuous hotspot study which was conducted in the City of Laredo to characterize levels of ultrafine particles and other pollutants for three different time periods. Dr. Kuruvilla John and his graduate students worked on a study that resulted in the development of a base case modeling system for the near non-attainment areas of Austin, San Antonio, Corpus Christi, and Victoria. The modeling system studied the impact of various emissions categories and the influence of climate change induced atmospheric warming on ozone air quality in each of the cities.

In the South Texas area along the Rio Grande, cities and industries are sparse to non-existent until reaching the area just west of Reynosa and Roma. From this point eastward to where the Rio Grande runs into the Gulf of Mexico, the population has increased significantly until it appears to be one city stretched along the river on the Texas side and on the Mexico side the cities center around the locations of the international bridges. Both industry and populations have raised the amount of pollution that becomes airborne and contaminates run-off from rain. Surface water quality is also affected by the treated outflow from waste water treatment plants along the Rio Grande and all the rivers serving the region. In coastal areas the waste water treatment plants' outflow is released into creeks that feed into marshes, lagoons, and bays.

Since the water withdrawals from the Rio Grande by industry, agriculture, and municipalities from its source in Colorado and the inflow from the feeder rivers, creeks, and arroyos on each side of the river affect the amount and quality of the water when it reaches its southern section, it is important to study the flow and quality as it completes its journey. Environmental restrictions on the Mexico side of the border are not the same as on the Texas side for air and water pollution. As a result, the Texas side must focus on the quantity and quality of the flows from the feeder creeks and arroyos that transport water from springs and rain run-off. This also applies to the Gulf Coast side of the region as water flows into the fresh water-salt water marshes, lagoons, and bays.

The research studies included herein focus on the Rio Grande, its feeder creeks, and the flows into the intercoastal waterways, marshes, lagoons, and bays. Dr. David Ramirez worked with his graduate students to develop absorbents for environmental uses. Activated carbons are widely used to remove contaminants from waste streams. His research focused on the production and physical characterization of carbonaceous adsorbents from low cost mesquite woodchips available in the South Texas region.

Dr. Jianhong (Jennifer) Ren and her graduate students focused on a semi-arid creek tributary of the Rio Grande near Laredo, Texas. The first study collected samples from six sites along the creek during the different seasons and analyzed for dissolved and total metals in water and total metals in sediments. The second study was an initial water and sediment quality study conducted at a site next to a decommissioned antimony plant on Manadas Creek, a tributary of the Rio Grande located in the Laredo area.

Dr. Lee W. Clapp and his graduate student studied the major environmental impacts of concentrate discharge on receiving waters downstream of the Southmost Regional Water Authority (SRWA) brackish groundwater desalination plant, which is located close to the Gulf Coast near Brownsville, Texas. Dr. Kim Jones and his graduate students focused on wetlands in the Lower Rio Grande Valley. The first study evaluated the hydrological conditions in a pilot- scale constructed channel wetland in South Texas. The second study evaluated (1) the relative water quality treatment ability for several wetland areas and (2) the recovery of nutrient removal activity after major flooding damage and inundation.

Acknowledgements This material is based upon work supported by the Center of Research Excellence in Science and Technology – Research on Environmental Sustainability of Semi-Arid Coastal Areas (CREST-RESSACA) at Texas A&M University– Kingsville (TAMUK) through a Cooperative Agreement (No. HRD-0206259) from the National Science Foundation (NSF). Any opinions, findings, and conclusions or recommendations expressed in this material are those of the author(s) and do not necessarily reflect the views of the National Science Foundation.

References

Chihuahuan Desert. (2001). *World wildlife organization*. Retrieved on April 8, 2011, from http://www.worldwildlife.org/wildworld/profiles/terrestrial/ma/na1301_full.html

Coahuila. (n.d.). *Wikipedia*. Retrieved on January 23, 2012, from http://en.wikipedia.org/wiki/Coahuila

Coahuila. Retrieved on January 23, 2012, from http://www.nationsencyclopedia.com/mexico/Aguascalientes-M-xico/Coahuila.html

Cook, T., Adams, J., Valero, A., Schipper, J., & Allnutt, T. (2001). Tamaulipan mezquital (NA1312). In *Wild world ecoregion profile*. World Wildlife Fund. Retrieved on April 8, 2011, from http://www.worldwildlife.org/wildworld/profiles/terrestrial/na/na1312_full.html

Eagle Ford Shale Trend Map. (n.d.). Retrieved on April 12, 2011, from http://www.petroresourcescorp.com/images/ops_texasgulflarge.jpg

Garner, L. E. (2012). Mineral resources and mining. In *The handbook of Texas online*. Retrieved on April 8, 2011, from http://www.tshaonline.org/handbook/online/articles/MM/gpm01.html

Hamilton, T. (2011, July 3). Drought spurring fracking concerns: Necessary water may be problem. *San Antonio Express News*, pp. 1B, 12B.

Hoekstra, J., Molnar, J., Jennings, M., Revenga, C., Spalding, M., Boucher, T., et al. (2010). *The atlas of global conservation: Changes, challenges, and opportunities to make a difference*. Berkley, CA: University of California Press. http://www.nature.org/ourscience/sciencefeatures/conservation-atlas-1.xml

Mertz, L. Rio Grande. *Handbook of Texas online*. Retrieved on February 6, 2012, from http://www.tshaonline.org/handbook/online/articles/rnr05. Published by the Texas State Historical Association.

Olson, D. M., Dinerstein, E., Wikramanayake, E. D., Burgess, N. D., Powell, G. V. N., Underwood, E. C., et al. (2001). Terrestrial ecoregions of the world: A new map of life on earth. *BioScience, 51*, 933–938. More information at: www.worldwildlife.org/science/ecoregions/item1267.html. Ecoregion map found at http://www.nature.org/ourscience/sciencefeatures/conservation-atlas-1.xml

Rand McNally & Company. (2000). *Road atlas of the United States, Canada, Mexico*. Skokie, IL: Rand McNally & Co.

Rio Grande. (n.d.). *Wikipedia*. Retrieved on April 8, 2011, from http://en.wikipedia.org/wiki/Rio_Grande

Rio Grande Valley. (n.d.). *Wikipedia*. Retrieved on Aril 12, 2011, from http://en.wikipedia.org/wiki/Rio_Grande_Valley

South Texas. (n.d.). *Wikipedia*. Retrieved on April 11, 2011, from http://en.wikipedia.org/wiki/South_Texas

Spearing, D. (1991). *Roadside geology of Texas*. Missoula, MT: Mountain Press.

Tamaulipan mezquital. (n.d.). *Terrestrial ecoregions*. World Wildlife Fund. Retrieved on April 11, 2011, from http://worldwildlife.org/ecoregions/na1312

Texas Parks and Wildlife Department [TPWD]. (2008a). Habitats. In *Land and water topics*. Retrieved on April 8, 2011, from http://www.tpwd.state.tx.us/landwater/land/habitats/

TPWD. (2008b). South Texas plains wildlife district. In *Land and water topics*. Retrieved on April 8, 2011, from http://www.tpwd.state.tx.us/landwater/land/habitats/southtx_plain/regulatory/

United States Census Bureau. (2010). *American fact finder: 2010 census*. Retrieved on October 31, 2012, from http://www.census.gov/popest/data/cities/totals/2011/index.html, http://www.census.gov/popest/data/cities/totals/2011/files/SUB-EST2011_48.csv

Chapter 2
Pollution and Lifestyle Causes of Asthma and Allergies Among School Children of Tamaulipas, Mexico in the U.S.–Mexico Border Region

Laura Guerrero-Medrano, Prajay Gor, Ruth Reyna-Caamaño, Gerardo M. Mejia-Velazquez, Jesus Santos-Guzman, and Alvaro Martinez

Abstract The prevalence of asthma, allergic rhinitis, and eczema in school children in the cities of Northeastern Mexican State of Tamaulipas and potential correlation with ambient pollution factors were studied using the International Study of Asthma and Allergies in Childhood (ISAAC) epidemiology research methodology Phase 1. The study indicates that asthma and rhinitis are more prevalent than eczema in both elementary and middle school children with rhinitis prevalence significantly higher in Reynosa elementary school children. Asthma was more prevalent in Nuevo Laredo middle school children than in the other cities' middle school children likely due to higher frequency of smoking among children as a very strong correlation was found between asthma occurrence and children smoking habits. Lead and pesticide intoxications are higher in middle school children than in elementary school children. Middle school children in Nuevo Laredo exhibit two or more time the pesticide intoxications than middle school children in Reynosa and Matamoros; however, the study indicates that Matamoros

L. Guerrero-Medrano • J. Santos-Guzman
Instituto Tecnológico y de Estudios Superiores de Monterrey (ITESM), Eugenio Garza Sada 2501, Monterrey, Nuevo León CP 64849, Mexico

School of Medicine and Health Sciences, Tecnologico de Monterrey, Monterrey, Nuevo León, Mexico

P. Gor • A. Martinez (✉)
Department of Environmental Engineering, Frank H. Dotterweich College of Engineering, Texas A&M University-Kingsville, 700 University Blvd. MSC 213, Kingsville, TX 78363, USA
e-mail: alvaro.martinez@tamuk.edu

R. Reyna-Caamaño • G.M. Mejia-Velazquez
Instituto Tecnológico y de Estudios Superiores de Monterrey (ITESM), Eugenio Garza Sada 2501, Monterrey, Nuevo León CP 64849, Mexico

Center for Environmental Quality, School of Engineering and Information Technologies, Tecnologico de Monterrey, Monterrey, Nuevo León, Mexico

D. Ramirez et al. (eds.), *Environmental Sustainability Issues in the South Texas–Mexico Border Region*, DOI 10.1007/978-94-007-7122-2_2,

has the highest outdoor fumigation and Nuevo Laredo the lowest indoor fumigation. Indoor air quality may also be a factor as the study shows the combined results for the three cities of Matamoros, Reynosa, and Nuevo Laredo that 60 % of households use coal or wood for heating purposes, and 60 % of households use natural gas for cooking.

Keywords Air quality • Environmental health indicators • Health effects • Mortality cases • Mortality rates • South Texas-Northeastern Mexico border region

2.1 Introduction

The United States (U.S.)–Mexico border region have grown rapidly in the last few decades due to improved economy, rapid industrialization, greater trans-border linkages, and growth in population (Pick, Viswanathan, & Hettrick, 2001). The border population grew from 7 million in 1980 to 13 million in 2005. Since the implementation of the North American Free Trade Agreement (NAFTA) in 1993, the border regions have experienced an expansion of industrial complexes known as *maquiladoras* (Shi, Fernando, & Yang, 2009). The section of the border region between South Texas and Northeastern Mexico is an area of special interest since large cities, *colonias* and industries are located on both sides of the border (Mejia-Velazquez & Rodriguez-Gallegos, 1997). Rapid economic growth and increasing population in this border region have led to several environmental and health concerns.

Air pollution is a major environmental concern. Studies have suggested that air pollutants cause various adverse health effects. The effects of environmental air pollution are potentially more adverse on sensitive population such as children and elderly and make them. more vulnerable and susceptible to environmental risk (International Study of Asthma and Allergies in Childhood [ISAAC], 1993; Mejía, Choy, Mengersen, & Morawska, 2011; Samoli, Nastos, Paliatsos, Katsouyanni, & Priftis, 2011). Children seem to be the most vulnerable to air pollution because their defense mechanism is not fully developed and they inhale a higher volume of air per body weight than adults. In addition they perform a greater level of physical activity and spend more time outdoors than adults (Salvi, 2007; Tabaku, Bejtja, Bala, Toci, & Resuli, 2011). At school children are exposed to emissions from sources within and outside the school. Studies have shown higher prevalence of illness among children attending schools near industrial and mobile sources (Mejía et al., 2011). The observed illnesses from air pollution in children include increased respiratory symptoms and diseases, aggravation of asthma, decrease in lung function, and school absenteeism (He et al., 2010). These effects have led to high health service costs in admissions and hospitalizations (Feitosa et al., 2011). Environmental health effects are commonly associated with smoking, air pollution, pesticides, and other toxic chemicals which have been identified as key unfavorable environmental pollutants by the United States Environmental Protection Agency (U.S. Environmental Protection Agency [U.S. EPA], 2006). Environmental health effects of asthma and allergies such as rhinitis, dermatitis, and eczema among children have been increasing globally for the past few decades (Leung, Ko, & Wong 2012; Peñaranda, Aristizabal, García, Vásquez, & Rodríguez-Martinez, 2012). This increasing trend can be attributed to changes associated with environmental and lifestyle changes during this modernization period. Indoor air pollution caused from the use of wood or coal burning for cooking or heating purposes can also cause significant particle pollution (Leung et al., 2012). The environmental health effects of asthma, rhinitis, and eczema, particularly in children, in the border region have not been studied sufficiently. ISAAC is a worldwide epidemiological research program established to study prevalence, severity, and variations in childhood asthma and allergies such as rhinitis and eczema at the population level, comparison of these diseases worldwide, monitor and obtain baseline measures for assessment of future trends, and provide a framework for further etiological research into genetic, lifestyle, environmental, and medical care factors affecting these diseases (ISAAC; Peñaranda et al., 2012). ISAAC was the first investigation carried out worldwide using standardized written and video questionnaires to study and generate a global map of childhood asthma and allergies such as rhinitis and eczema and to monitor future trends in these diseases (Behbehani et al., 2000; Pegas et al., 2011).

The present study is focused on determining the prevalence of asthma, rhinitis, and eczema in school children by applying a modified ISAAC questionnaire for the cities of Matamoros, Reynosa, and Nuevo Laredo in the State of Tamaulipas in Northeastern Mexico. Several studies have demonstrated that children can be screened for asthma and allergies by applying the ISAAC questionnaire at both the elementary and middle school ages (Peroni, Piacentini, Bodini, & Boner, 2009). The populations of Tamaulipas have grown from about 720,000 in 1950 to 2,750,000 in 2000, which is about a 400 % increase for that period (Pick et al., 2001). The ISAAC questionnaire methodology was chosen for this study because environmental and health information is incomplete and not readily available for this region. Three cities in the state of Tamaulipas were selected for investigation, Matamoros, Reynosa, and Nuevo Laredo. Table 2.1 summarizes the population growth and Fig. 2.1 shows the location of these cities within the South Texas-Tamaulipas border region.

The increase in the population of Tamaulipas can be attributed to domestic migration due to the relative economic prosperity resulting from the increasing U.S.–Mexico commercial flow and the jobs created by the *maquiladoras* industry. These cities also serve as holding zones for migration to the United States (Pick et al., 2001).

The research aimed to determine the prevalence of asthma, rhinitis, and eczema in children living in the selected cities located in Tamaulipas, determine the potential causes, and draw conclusions from comparisons between schools. This research also intended to establish a baseline for assessment of future trends in the prevalence of these diseases and to provide a framework for further epidemiological research of factors affecting these diseases such as genetics, lifestyle, environment, and medical care.

Table 2.1 Population growth of Tamaulipas cities, 1930–2020 (Pick et al., 2001)

Cities	1930	1950	1970	1990	2020 projected	Annual growth rate
Matamoros	9,733	45,846	140,660	303,295	736,891	4.8
Reynosa	4,840	34,087	140,480	282,666	658,403	5.5
Nuevo Laredo	21,636	57,668	152,325	219,465	633,770	3.8

Fig. 2.1 Location of Tamaulipas cities in the Northeastern Mexico-Texas border region

2.2 Methodology

2.2.1 ISAAC Study Phases

The ISAAC study typically has three phases as described in Table 2.2 (ISAAC, 1993).

2.2.2 Selection of the Study Population

This study concentrated on Phase I of the ISAAC study design phases. Phase I is a compulsory core study designed to assess the prevalence and severity of asthma and allergic diseases in the defined population. The population in this study consisted of approximately 3,000 children aged 6–7 and 13–14 years old attending elementary and middle schools in the three border cities of Matamoros, Reynosa, and Nuevo Laredo in the Mexican State of Tamaulipas. Children participating in the survey were selected at random from those meeting the age criterion identified from the

Table 2.2 ISAAC study design phases

Phase I	A compulsory core study designed to assess the prevalence and severity of asthma and allergic disease in defined populations
Phase II	Study will be developed to investigate possible etiological factors, particularly those suggested by the findings of Phase I, and involves more detailed studies of etiological factors and clinical examination of subgroups of children
Phase III	A repetition of Phase I will be initiated after a period of 3 years

schools' class registers. It is recognized that there may be some children outside the specified age ranges in each class chosen. These children will be included in the data collection, but will be excluded from analysis for the international comparison. The younger age group (6–7 years of age) was chosen to give a reflection of the early childhood years when asthma is common and admission rates are particularly higher. The older age group (13–14 years of age) was chosen to reflect the period when mortality from asthma is more common. As explained in the next section, the parents or the children themselves were asked to complete the ISAAC questionnaire on asthma, rhinitis, and eczema.

The schools selected for the study represent major city sectors and proximity to existing risk factors such as industrial parks, open landfills, pollution emission point sources, and high-traffic roads with high mobile source emissions. In each city 12 elementary and 12 middle schools were selected. It was the intent to distribute 2,400 questionnaires per city, with the goal of collecting 30 %. Wherever possible, the schools selected had equivalent children populations.

2.2.3 Legal, Ethical, and Logistic Considerations

With the Public School Secretary's authorization, private and public school directors were contacted for individual authorization. Once the directors gave their approval to the project, the children parents were asked for their consent to respond to the questionnaire or for their child to respond to the questionnaire. For the elementary school children (6–7 years of age), informative parental meetings were held in elementary schools to provide them full disclosure related to the project, including how to complete the ISAAC questionnaire by fully explaining the associated set of instructions. The parents answered the questionnaires at home and returned them to the school. For the middle school children (13–14 years of age), the ISAAC questionnaire was answered by them in their school under the supervision of their classroom teacher. Prior to answering the questionnaires the middle school children received a brief informative lecture on respiratory diseases and their preventive measures.

2.2.4 ISAAC Self-Answered Questionnaire Data

The prevalence of diseases among school children was determined through the modified ISAAC questionnaire, intended for parents of 6–7 year old elementary school children and for 13–14 year old middle school children themselves. Both questionnaires had identical questions, the only difference was that one of the questionnaires was answered by the parents of the elementary school children and the other questionnaire was answered by the middle school children themselves. The modification of the original ISAAC questionnaire consisted of additional questions asking for chemical intoxications such as lead intoxication due to use of paints at home, pesticide intoxication due to use of pesticides and use of indoor insecticide aerosols and burn-sticks, indoor and outdoor fumigation in the vicinity of the home, and type of indoor combustion sources used for heating and cooking purposes. No lead intoxication or lead paint use confirmation was performed; only the self-answered data from questionnaire was used for evaluation. The questionnaire also gathered other information, such as children's place of birth, personal health, and indoor and outdoor risk factors.

2.2.5 Data Analysis

The questionnaire recovery efficiency was determined from the total number of distributed questionnaire and the ones that were returned fully answered. Only the returned fully answered questionnaires were used for data analysis.

2.3 Results and Discussion

Data recovery varied between elementary and middle schools, but remained constant between cities. The elementary school children's questionnaire was sent to their homes for completion and had a recovery rate of 30 %. The middle school children questionnaire were distributed and completed in the classroom during the school day and had a recovery rate of 90 %.

Figures 2.2, 2.3, and 2.4 show maps of Nuevo Laredo, Reynosa, and Matamoros, respectively, subdivided into areas and zones. The figures also indicate school and industry locations.

Table 2.3 indicates the number of ISAAC questionnaire collected from the elementary and middle schools by city.

In this epidemiological study, prevalence was determined as a measure of the proportion of children in schools who had one or more of the studied diseases at the particular time the ISAAC questionnaire was administered. Table 2.4 shows the percentage of children reporting respiratory diseases such as asthma, rhinitis,

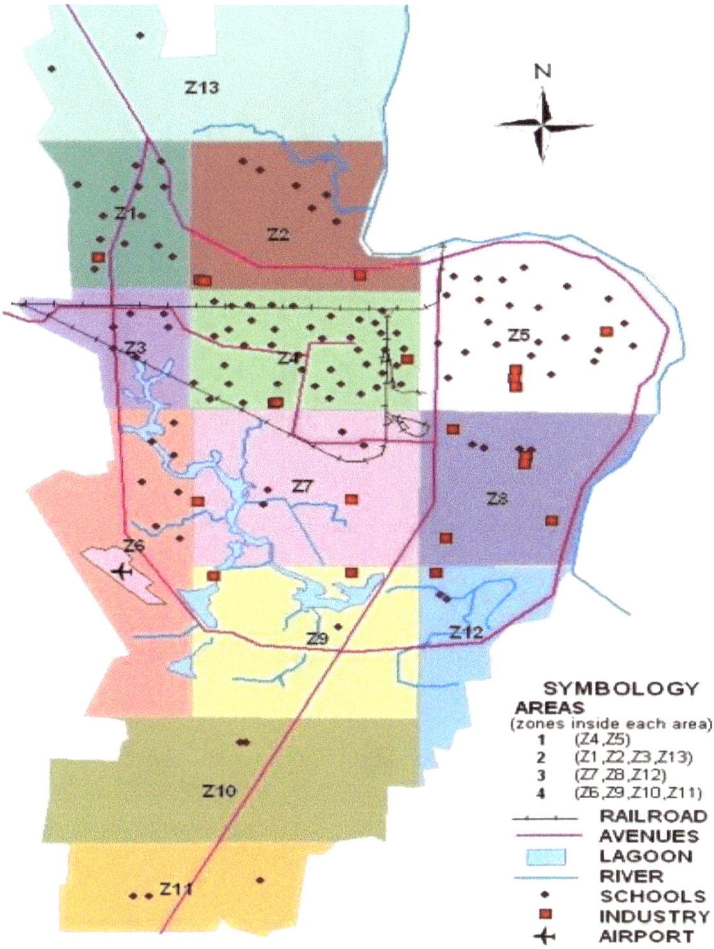

Fig. 2.2 Nuevo Laredo areas and zones with school and industry locations

and eczema on their ISAAC questionnaire. Table 2.5 shows the number of cases of respiratory diseases by age group.

Figure 2.5 shows the prevalence values of various respiratory diseases such as asthma, rhinitis, and eczema by age group for the selected Tamaulipas cities.

The data indicates that the prevalence of asthma and rhinitis is higher than eczema in both elementary and middle school children. Asthma was more prevalent in middle school children in Nuevo Laredo and Reynosa while more prevalent in elementary school children of Matamoros. Rhinitis was more prevalent in elementary school children in Reynosa and Matamoros. In Nuevo Laredo it was more prevalent in middle school children. The prevalence of rhinitis was highest in Reynosa's elementary school children. Eczema appeared to be more prevalent in elementary school children than in middle school children.

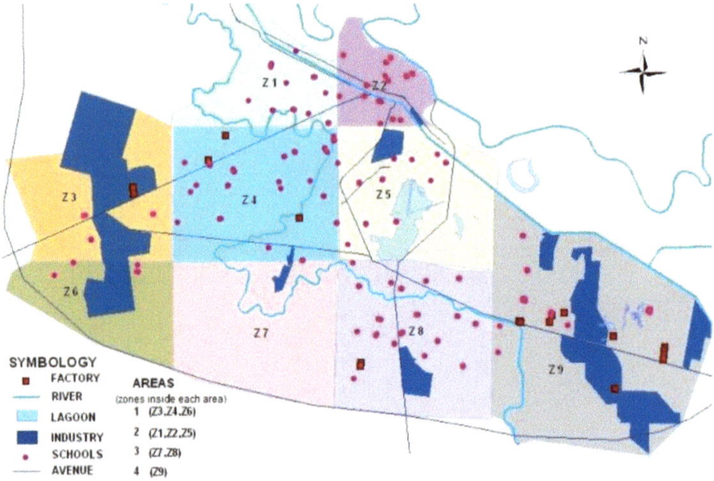

Fig. 2.3 Reynosa areas and zones with school and industry locations

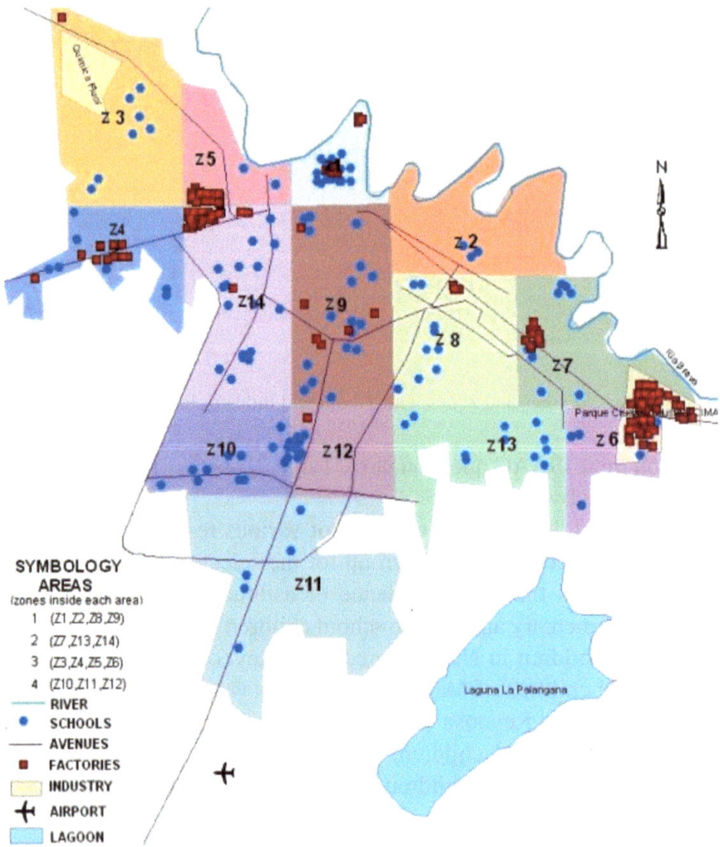

Fig. 2.4 Matamoros areas and zones with school and industry locations

Table 2.3 Total number of ISAAC questionnaire collected

City	Elementary school	Middle school	Total
Nuevo Laredo	155	629	784
Reynosa	197	1,145	1,342
Matamoros	275	1,015	1,290
Total	627	2,789	3,416

Table 2.4 Children survey percentage results by respiratory disease

	% reporting	Without disease	With disease	Do not know
Total	Asthma	93.82	6.18	
	Rhinitis	64.89	8.23	26.87
	Eczema	96.49	3.51	
Nuevo Laredo	Asthma	91.45	8.55	
	Rhinitis	64.96	7.93	27.11
	Eczema	96.3	3.7	
Reynosa	Asthma	93.88	6.12	
	Rhinitis	64.42	8.97	26.61
	Eczema	95.97	4.03	
Matamoros	Asthma	95.19	4.81	
	Rhinitis	65.34	7.65	27.01
	Eczema	97.13	2.87	

Table 2.5 Number of cases by respiratory disease by age group and city

	Nuevo Laredo		Reynosa		Matamoros	
	6–7 years of age	13–14 years of age	6–7 years of age	13–14 years of age	6–7 years of age	13–14 years of age
Asthma	11	56	9	73	19	43
Rhinitis	65	215	28	92	27	71
Eczema	8	21	13	41	9	28

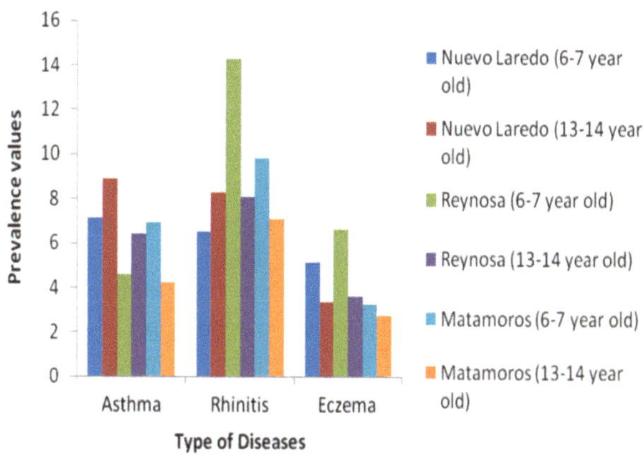

Fig. 2.5 Prevalence values of respiratory and allergic diseases by age group and selected city

Table 2.6 Children (13–14 years of age) smoking habits from ISAAC questionnaire

Number of cigarettes/day	Total (%)	Nuevo Laredo (%)	Reynosa (%)	Matamoros (%)
1	3.96	6.38	3.69	2.5
1–20	1.12	1.52	1.14	0.83
>20	0.69	0.61	0.57	0.83

Table 2.7 Percentage of coughing after tobacco smoke exposure

	Total	Nuevo Laredo	Reynosa	Matamoros
No	42.69	43.96	44.34	40.28
Sometimes	28.18	32.97	24.21	28.06
Yes	29.13	23.08	31.45	31.67

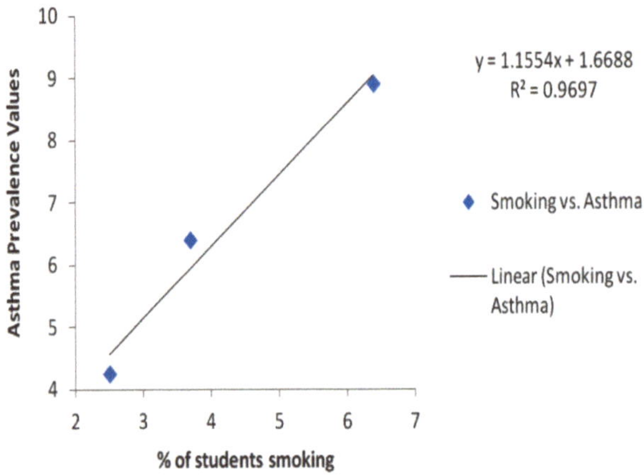

Fig. 2.6 Correlation between cigarette smoking and asthma prevalence

The middle school children's ISAAC questionnaire also explored the risk of respiratory diseases associated with cigarette smoking among children (13–14 years of age). Table 2.6 shows the children smoking habits for the cities of Nuevo Laredo, Reynosa, and Matamoros. The table indicates that most children smokers have only one cigarette a day. It also indicates that children in the middle schools of Nuevo Laredo reported more tobacco use than in Reynosa and Matamoros.

Table 2.7 shows the percentage of children coughing as a tobacco smoke side effect. Data indicated that more than 50 % of the middle school children reported coughing after tobacco smoke exposure for all the selected Tamaulipas cities.

Figure 2.6 shows the correlation between the percentage of children smoking one cigarette a day and asthma prevalence among children in the middle school based on ISAAC questionnaire data.

A very strong correlation ($R^2 = 0.9697$) between cigarette smoking and asthma prevalence among middle school children (13–14 years of age) was indicated by these data. Since a higher percentage of middle school children in Nuevo Laredo smoke, this could be the reason for the high asthma prevalence amongst these children.

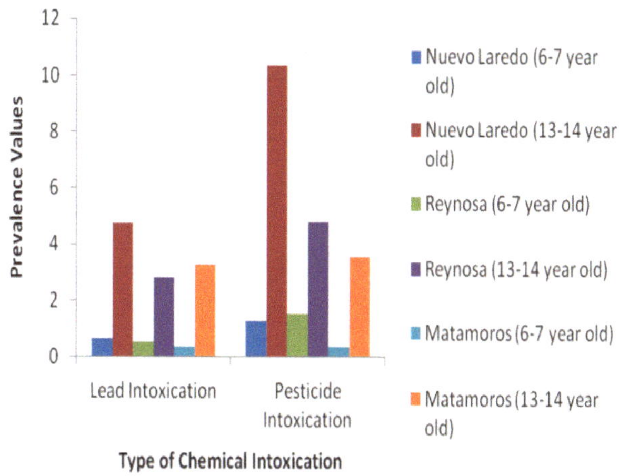

Fig. 2.7 Prevalence of chemical intoxication due to lead and pesticide use

Table 2.8 Prevalence of fumigation by cities

	Nuevo Laredo		Reynosa		Matamoros	
Fumigation	Outdoor	Indoor	Outdoor	Indoor	Outdoor	Indoor
Once a week	33.75	37.38	44.21	32.93	42.18	34.07
2–3/week	39.12	49.22	27.3	46.34	17.52	47.35
Daily	27.13	13.4	28.49	20.73	40.52	18.58

Figure 2.7 shows prevalence of self reported chemical intoxication among elementary and middle school children in the selected Tamaulipas cities due to lead and pesticide use. As previously noted, parents of elementary school children (6–7 years of age) completed the ISAAC questionnaire, while the middle school students (13–14 years of age) completed the form themselves.

It was observed that lead and pesticide intoxication was higher in middle school children (13–14 years of age) than in elementary school children (6–7 years of age) for all three cities. Middle school children in Nuevo Laredo showed two or more times the pesticide intoxication when compared to the middle school children in Reynosa and Matamoros.

Table 2.8 shows prevalence of outdoor and indoor fumigation for Nuevo Laredo, Reynosa, and Matamoros obtained from the ISAAC questionnaire data. On a daily basis, outdoor fumigation was more common than indoor fumigation, but the opposite was reported for the 2–3 times of fumigation per week for the selected Tamaulipas cities. Matamoros reported the highest outdoor fumigation on a daily basis, and Nuevo Laredo reported lowest indoor fumigation on a daily basis.

Table 2.9 shows the prevalence of different combustion sources used indoors in the cities of Nuevo Laredo, Reynosa, and Matamoros. The combined results for the three cities shows 60 % of households use coal or wood for heating purposes, and 60 % of households use natural gas for cooking purposes.

Table 2.9 Indoor combustion sources by cities, based on ISAAC questionnaire

	Fuel use	Natural gas	LP gas	Coal/wood	Other
Nuevo Laredo	Cooking	34.26	39.2	23.46	3.09
	Water heater	1.83	39.14	58.41	0.61
Reynosa	Cooking	36.08	29.26	29.55	5.11
	Heater	1.44	20.69	77.3	0.57
Matamoros	Cooking	28.57	37.01	30.52	3.9
	Water heater	1.43	26.53	71.43	0.61

2.4 Conclusions

This study addressed the prevalence of the environmental health effects of asthma, rhinitis, and eczema, specifically in the school children residing in three cities located in the Northeastern Mexican State of Tamaulipas located along the South Texas border region. The study was based on the ISAAC questionnaire methodology Phase I since there was a scarcity of health statistic data readily available for Tamaulipas. The resulting outcomes were based on the ISAAC questionnaires answered by the parents of elementary school children (6–7 years of age) and by the middle school children (13–14 years of age) in the cities of Nuevo Laredo, Reynosa, and Matamoros.

Results of the ISAAC questionnaire data indicate that asthma and rhinitis were more prevalent than eczema in both elementary and middle school children for the selected Tamaulipas cities. Rhinitis prevalence was significantly higher in Reynosa elementary school children than any other schools for all cities. Asthma was more prevalent in Nuevo Laredo middle school children than any of the middle schools in Reynosa and Matamoros. This could be due to the higher percentage of children smoking in middle schools of Nuevo Laredo since a very strong correlation ($R^2 = 0.9697$) between asthma prevalence values and percentage of children smoking was determined.

The data indicated that lead and pesticide intoxication was higher in middle school children than in elementary school children for all three cities. Middle school children in Nuevo Laredo showed two or more times the pesticide intoxication when compared to middle school children in Reynosa and Matamoros.

For all the cities in this study, the results indicated that outdoor fumigation was more common than indoor fumigation on a daily basis, but the opposite can be seen for the 2–3 times of fumigation per week. On a daily basis, Matamoros reported the highest outdoor fumigation while Nuevo Laredo reported lowest indoor fumigation. The data also indicated that the combined results for all three cities shows 60 % of households use coal or wood for heating purposes, and 60 % of households use natural gas for cooking purposes.

The results of the ISAAC questionnaire provided evidence for the prevalence of asthma, rhinitis, and eczema in the elementary and middle school children in the three cities located in Tamaulipas and the environmental health effects of pesticide fumigation and indoor combustion sources.

Acknowledgements This material is based upon work supported by the Center of Research Excellence in Science and Technology – Research on Environmental Sustainability of Semi-Arid Coastal Areas (CREST-RESSACA) at Texas A&M University– Kingsville (TAMUK) through a Cooperative Agreement (No. HRD-0206259) from the National Science Foundation (NSF). Any opinions, findings, and conclusions or recommendations expressed in this material are those of the author(s) and do not necessarily reflect the views of the National Science Foundation.

References

Behbehani, N. A., Abal, A., Syabbalo, N. C., Azeem, A. A., Shareef, E., & Al-Momen, J. (2000). Prevalence of asthma, allergic rhinitis, and eczema in 13- to 14-year-old children in Kuwait: An ISAAC study. *Annals of Allergy, Asthma & Immunology, 85*(1), 58–63.

Feitosa, C. A., Santos, D. N., Barreto do Carmo, M. B., Santos, L. M., Teles, C. A. S., Rodrigues, L. C., et al. (2011). Behavior problems and prevalence of asthma symptoms among Brazilian children. *Journal of Psychosomatic Research, 71*(3), 160–165.

He, Q.-Q., Wong, T. W., Du, L., Jiang, Z.-Q., Gao, Y., Qiu, H., et al. (2010). Effects of ambient air pollution on lung function growth in Chinese schoolchildren. *Respiratory Medicine, 104*(10), 1512–1520.

International Study of Asthma and Allergies in Childhood. (1993). *International study of asthma and allergies in childhood manual*. Auckland, NZ: ISAAC.

Leung, T. F., Ko, F. W.-S., & Wong, G. W.-K. (2012). Roles of pollution in the prevalence and exacerbations of allergic diseases in Asia. *The Journal of Allergy and Clinical Immunology, 129*(1), 42–47.

Mejía, J. F., Choy, S. L., Mengersen, K., & Morawska, L. (2011). Methodology for assessing exposure and impacts of air pollutants in school children: Data collection, analysis and health effects – A literature review. *Atmospheric Environment, 45*(4), 813–823.

Mejia-Velazquez, G., & Rodriguez-Gallegos, M. (1997). Characteristics and estimated air pollutant emissions from fuel burning by the industry and vehicles in the Matamoros – Reynosa border region. *Environment International, 23*(5), 733–744.

Pegas, P. N., Alves, C. A., Scotto, M. G., Evtyugina, M. G., Pio, C. A., & Freitas, M. C. (2011). Risk factors and prevalence of asthma and rhinitis among primary school children in Lisbon. *Revista Portuguesa de Pneumologia, 17*(3), 109–116 (English Edition).

Peñaranda, A., Aristizabal, G., García, E., Vásquez, C., & Rodríguez-Martinez, C. E. (2012). Rhinoconjunctivitis prevalence and associated factors in school children aged 6–7 and 13–14 years old in Bogota, Colombia. *International Journal of Pediatric Otorhinolaryngology, 76*(4), 530–535.

Peroni, D. G., Piacentini, G. L., Bodini, A., & Boner, A. L. (2009). Preschool asthma in Italy: Prevalence, risk factors and health resource utilization. *Respiratory Medicine, 103*(1), 104–108.

Pick, J., Viswanathan, N., & Hettrick, J. (2001). The U.S.–Mexican borderlands region: A binational spatial analysis. *The Social Science Journal, 38*, 567–595.

Salvi, S. (2007). Health effects of ambient air pollution in children. *Paediatric Respiratory Reviews, 8*(4), 275–280.

Samoli, E., Nastos, P. T., Paliatsos, A. G., Katsouyanni, K., & Priftis, K. N. (2011). Acute effects of air pollution on pediatric asthma exacerbation: Evidence of association and effect modification. *Environmental Research, 111*(3), 418–424.

Shi, C., Fernando, H., & Yang, J. (2009). Contributors to ozone episodes in three U.S.–Mexico border twin–cities. *Science of the Total Environment, 407*, 5128–5138.

Tabaku, A., Bejtja, G., Bala, S., Toci, E., & Resuli, J. (2011). Effects of air pollution on children's pulmonary health. *Atmospheric Environment, 45*(40), 7540–7545.

U.S. Environmental Protection Agency [U.S. EPA]. (2006). *Border 2012: U.S.–Mexico environment program indicators report 2005*. Washington, DC: Office of Environmental Information: EPA–160–R–06–001.

Chapter 3
Environmental Health Indicators for the Border Region Between South Texas and Northeastern Mexico

Prajay Gor, Alvaro Martinez, Ruth Reyna-Caamaño, Gerardo M. Mejia-Velazquez, Laura Guerrero-Medrano, and Jesus Santos-Guzman

Abstract The United States – Mexico border region have experienced high economic and population growth which have led to environmental consequences, such as air pollution, that affect human health. The objective of this study was to assess the impacts of air quality on public health for the South Texas – Northeastern Mexico border region by means of environmental health indicators that correlate air quality data with epidemiological data for respiratory and cardiovascular illness, and facilitates interpretation of outcomes and trends. The study established that particulate matter (PM), both PM_{10} and $PM_{2.5}$, have a significant effect on heart disease, stroke, and asthma in South Texas. The data show that asthma is prevalent along the border region, and for the Mexican State of Tamaulipas, the study demonstrated a higher correlation between PM_{10} and asthma than any other disease. The study also shows a statistically strong correlation between ambient ozone (O_3) concentrations and mortality due to heart disease. Although some correlation appears to exist between other air pollutants such as carbon monoxide (CO), O_3

P. Gor • A. Martinez (✉)
Department of Environmental Engineering, Frank H. Dotterweich College of Engineering, Texas A&M University-Kingsville, Kingsville, TX 78363, USA
e-mail: alvaro.martinez@tamuk.edu

R. Reyna-Caamaño • G.M. Mejia-Velazquez
Instituto Tecnológico y de Estudios Superiores de Monterrey (ITESM), Monterrey, Nuevo León CP 64849, Mexico

Center for Environmental Quality, School of Engineering and Information Technologies, Tecnologico de Monterrey, Monterrey, Nuevo León, Mexico

L. Guerrero-Medrano • J. Santos-Guzman
Instituto Tecnológico y de Estudios Superiores de Monterrey (ITESM), Monterrey, Nuevo León CP 64849, Mexico

School of Medicine and Health Sciences, Tecnologico de Monterrey, Monterrey, Nuevo León, Mexico

D. Ramirez et al. (eds.), *Environmental Sustainability Issues in the South Texas–Mexico Border Region*, DOI 10.1007/978-94-007-7122-2_3,

and $PM_{2.5}$ with mortality rates due to stroke, chronic lower respiratory disease (CLRD), and with morbidity rate due to asthma, the statistical significance of these correlations were not supported by the t-test results.

Keywords Air quality • Environmental health indicators • Health effects • Mortality cases • Mortality rates • South Texas–Northeastern Mexico border region

3.1 Introduction

The United States (U.S.)–Mexico border region have experienced rapid development since the early 1980s. Economic transformation, rapid industrial expansion, and substantial increase in trade between the United States and Mexico have led to population growth in the border region. The border population grew from about 7 million in 1980 to 13 million in 2005 (Shi, Fernando, & Yang, 2009). It is also projected that the U.S. – Mexico border population will grow to 16–25 million by 2030 according to United States Environmental Protection Agency (U.S. EPA) (United States Environmental Protection Agency [U.S. EPA], 2011). Rapid population growth along the border region has caused development of unplanned infrastructure and demands on use of land, water, energy and natural resources (Norman, Feller, & Phillip Guertin, 2009). These changes have led to environmental stresses, and air pollution have been identified as one of the key unfavorable environmental impacts by government entities in both countries (U.S. EPA, 2006). Air pollution is a major environmental concern that affects human health, vegetation, forest species, and natural ecosystems (Afroz, Hassan, & Ibrahim, 2003).

Most of the air pollution problems in the border airsheds result from anthropogenic activities. Ambient air is polluted by emissions from power plants, industrial facilities, agricultural operations, motor vehicles, dust from unpaved roads, open burning of trash, and many other sources (U.S. EPA, 2006). These emissions may further threaten the air quality and may impact the health of the population in the border region. Concerns regarding the health impacts caused due to traffic-related air pollution are also high for populations along the U.S.–Mexico border region (Raysoni et al., 2011). The health impacts caused due to air pollution have received much attention (Beamish, Osornio-Vargas, & Wine, 2011). Many studies have reported adverse health effects of ambient air pollution on human health (Lee, Wang, Lu, Lin, & Hwang, 2011). Air pollutants such as particulate matter (PM) have been linked to causing increases in both hospital admissions and mortality due to lung and cardiovascular disease (Beamish et al., 2011). The health impacts of air pollutant ozone (O_3) includes increased rates in hospital admissions, emergency visits and mortality, increase in severity of chronic respiratory problems and decrease in lung function (Yang et al., 2012). Studies have shown that O_3 and particulate matter, both $PM_{2.5}$ and PM_{10} caused adverse effects to human health. Adverse effects includes symptoms that moderately affect normal daily activities to premature deaths (Hall, Brajer, & Lurmann, 2010). Many evidences have

Fig. 3.1 Border study region between South Texas and Northeastern Mexico

suggested that air pollution contributes to respiratory and allergic diseases, including asthma and chronic obstructive pulmonary disease (Laumbach & Kipen, 2012). People with heart disease, stroke, lung cancer and chronic lower respiratory disease (CLRD) have a higher risk of premature death if they are exposed to poor environmental air quality. Thus, reliable data is necessary in order to develop targeted effective measures to prevent the excessive increase in pollutant emissions.

Respiratory and cardiovascular illnesses are generally associated with poor air quality, but the health impacts of air quality in the border region have not been studied sufficiently. It is hypothesized that these health effects are not evenly distributed geographically, but instead are likely distributed according to differential population densities, mobility patterns, concentrations of point sources, distribution of fugitive sources, and transportation corridors.

To explore this hypothesis, the study focused on air quality and population health impacts in a study area along the South Texas–Northeastern Mexico border region as shown in Fig. 3.1. Data covers counties and major cities of South Texas and the State of Tamaulipas in Northeastern Mexico. Air quality and epidemiological data in the study area were also collected for respiratory and cardiovascular illness. Data were transformed by developing, adapting, and calculating environmental health indicators that normalize incidences on a per-capita basis or per unit of pollutant concentration.

The results of this study should be of particular relevance to regulators and policy makers entrusted with managing air quality issues along the border region of South Texas and Northeastern Mexico, as well as to scientists and engineers involved in air quality management.

3.2 Methodology

To analyze the interactions between air quality and health effects for the border region of South Texas and Northeastern Mexico, temporal and geographically distributed air quality data and epidemiological data were collected and environmental health indicators were introduced. Air quality data for South Texas were collected from the Texas Commission on Environmental Quality (TCEQ) air monitoring sites website (Texas Commission on Environmental Quality [TCEQ], 2007). Air quality data for the Northeastern Mexican State of Tamaulipas were provided to the investigators by the Office of Gobierno de Tamaulipas through personal communication. For these analyses the air quality of the border cities was considered associated to the entire county in which the cities are located.

The U.S. EPA considers six criteria air pollutants for air quality based on the adverse health effects they may cause to human health and the environment. Those six criteria air pollutants are carbon monoxide (CO), ozone (O_3), particulate matter (PM) in the range sizes of 2.5 μm ($PM_{2.5}$) and 10 μm (PM_{10}), nitrogen dioxide (NO_2), sulfur dioxide (SO_2), and lead (Pb). For this study only CO, O_3, $PM_{2.5}$, and PM_{10} were considered because the most complete data available at the air monitoring sites for this study area were only for those pollutants.

The major source of information for statistical health data for South Texas was the Texas Department of State Health Services (TDSHS) (Texas Department of State Health Services [TDSHS], 2007). Data for Tamaulipas border cities was provided to the investigators by the Office of the Secretaria de Salud del Estado de Tamaulipas through personal communication. Health statistics data expressed in terms of mortality/morbidity cases and mortality/morbidity rates for cardiovascular diseases such as heart disease and stroke and for respiratory diseases such as CLRD, lung cancer, and asthma were collected.

Comparing South Texas and Tamaulipas health data along with developing environmental indicators presented several challenges from the derived data gaps and from the different approaches used by agencies of the two countries to present the geographical distribution of the data. For instance, the Tamaulipas health data for different types of diseases, years, and geographic level sometimes were not fully available to the public, and the South Texas health data were only available at the county level by type of disease. Tamaulipas health data for its major cities have several levels of aggregation, temporality, and compatibility. Another problem was lack of completeness since the Office of Secretaria de Salud del Estado de Tamaulipas, the agency that provides Tamaulipas' mortality data, services only 39.1 % of the state's population.

Mortality data for all diseases in South Texas were collected from TDSHS from 1995 to 2003. Asthma morbidity cases for this study include data from most South Texas hospitals. Mortality data for the cities of Nuevo Laredo, Reynosa, and Matamoros in Tamaulipas were obtained from the Office of Secretaria de Salud del Estado de Tamaulipas for the years 2001–2006.

All mortality and morbidity statistics follow the *International Statistical Classification of Diseases and Related Health Problems, 10th Revision*. Age adjustments

Table 3.1 Cities in the study area

City	County/municipality	State	Country
Brownsville	Cameron	Texas	USA
Laredo	Webb	Texas	USA
Edinburg	Hidalgo	Texas	USA
Mission	Hidalgo	Texas	USA
Nuevo Laredo	Nuevo Laredo	Tamaulipas	Mexico
Reynosa	Reynosa	Tamaulipas	Mexico
Matamoros	Matamoros	Tamaulipas	Mexico

for the mortality and morbidity statistics are based on the 2000 Standard Population for the years 1999–2003 and the 1940 Standard Population for the years 1995–1998 (TDSHS, 2007).

Table 3.1 lists the cities comprising the study area and indicates the corresponding county, state, and country.

Environmental health indicators (EHIs) provide linkages between environmental parameters and human health based on scientific evidence and targets the development of specific management policies and actions based on their interpretation (Briggs, Corvalan, & Nurminen, 1996; Kjellstrom & Corvalan, 1995). EHIs make environmental and health data more manageable in order to provide quantitative measures of environmental conditions linked with health outcomes. EHIs can help in identifying the trends, monitor progress and establish priorities related to environmental conditions and health outcomes and thus help decision-makers in formation and implementation of environmental policy (Bell et al., 2011). EHIs are proposed to assess the relationship between environmental air quality and health effects for the South Texas – Northeastern Mexico border region. Each indicator is measured as a ratio of morbidity or mortality rates, measured in units of cases per 100,000 people, to environmental burden, measured in units of concentration of pollutant of interest. EHIs are calculated by the formula

$$EHI_i = \frac{[Mortality\,or\,MorbidityRate]_i}{[Environmental\,Burden]_i}, \tag{3.1}$$

where sub index i refers to the conventional pollutant of concern, such as PM_{10}.

To further assess if there were statistical correlations between mortality and morbidity rates and pollutant concentrations, scattered plots were presented and the statistical significance was determined using t-tests.

3.3 Results and Discussion

3.3.1 Air Quality Data

Air quality data for the criteria air pollutants CO, O_3, $PM_{2.5}$, and PM_{10} for the South Texas major cities of Brownsville, Laredo, Edinburg, and Mission was obtained

Table 3.2 CAMS locations in South Texas

City	County	CAMS	EPA site number
Brownsville	Cameron	C80/AGP180	48-061-0006
Laredo	Webb	C44/AFGP144/P244	48-479-0016
Edinburg	Hidalgo	C42/AP142	48-215-0042
Mission	Hidalgo	C43/AP143	48-215-0043

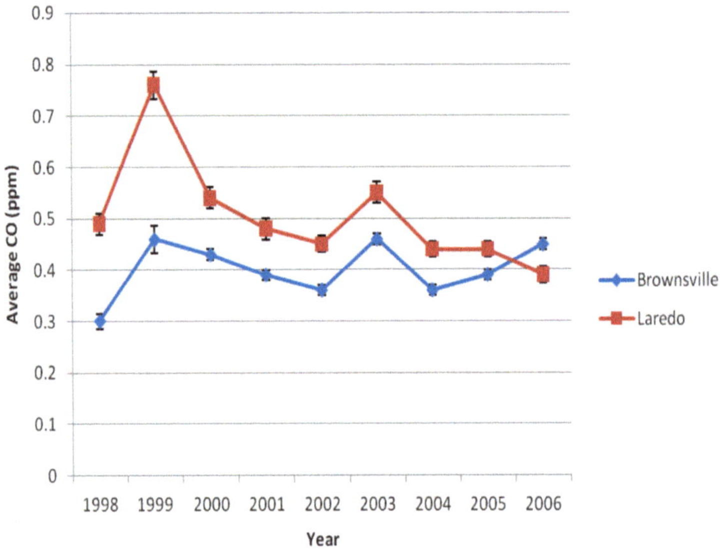

Fig. 3.2 Annual average carbon monoxide concentrations for Brownsville, TX and Laredo, TX

from the TCEQ website, which is gathered at their Continuous Air Monitoring Stations (CAMS). Table 3.2 shows the CAMS locations in the study region.

Figure 3.2 shows annual average concentrations and standard error bars for CO in the South Texas border cities of Brownsville and Laredo (TCEQ, 2007). The standard error bars represents the standard error of the mean, that is, annual average air pollutant concentration, which is estimated by its standard deviation divided by the square root of the sample size which, in this case, is 365 days of measurement in a year.

The annual average CO concentrations for Laredo and Brownsville are below the National Ambient Air Quality Standard (NAAQS) of 9 parts per million (ppm) for both cities. Laredo had exhibited relatively higher CO concentrations with a decreasing trend than Brownsville, and the concentration levels in 2006 were, for the first time, slightly lower than at Brownsville. Carbon monoxide (CO) emissions are commonly associated with mobile sources and higher concentrations may be associated with more vehicles and border crossing traffic through the international bridges. No CO data was available for the State of Tamaulipas border cities.

Figure 3.3 shows annual average ozone (O_3) concentration with standard error bars for the South Texas border cities of Brownsville, Laredo, Edinburg, and Mission.

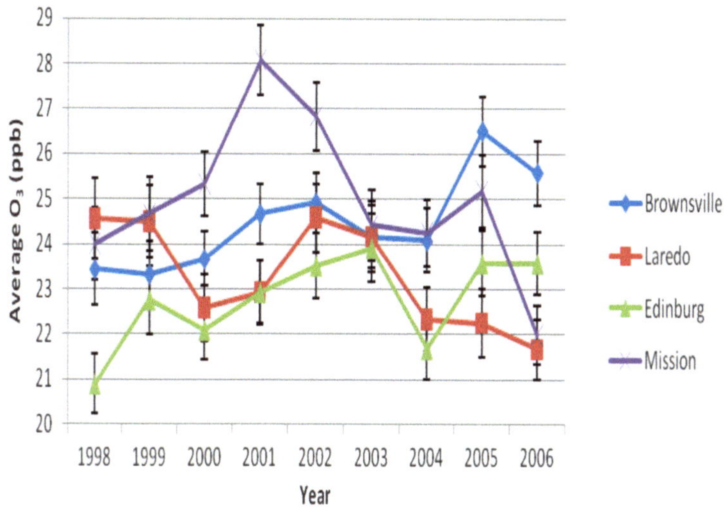

Fig. 3.3 Annual average ozone concentrations for South Texas border cities

Fig. 3.4 Annual average $PM_{2.5}$ concentrations for Brownsville and Mission

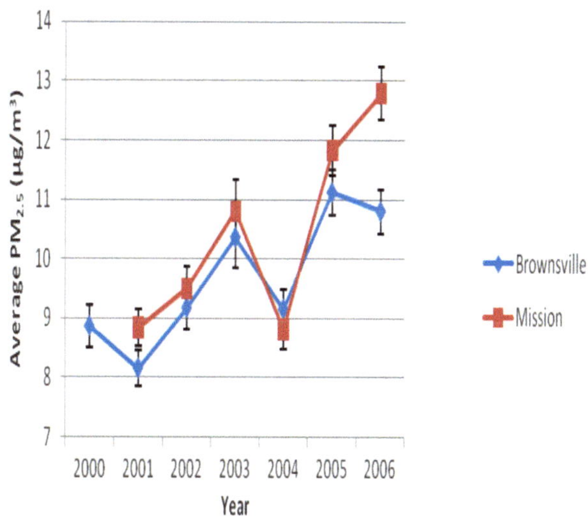

Ozone (O_3) levels for all the cities are below the NAAQS standard of 75 parts per billion (ppb). It is apparent that O_3 concentrations have been decreasing for the cities of Laredo and Mission, whereas they have been increasing for the cities of Brownsville and Edinburg. No O_3 data was available from Tamaulipas border cities.

Figure 3.4 shows annual average $PM_{2.5}$ concentrations and standard error bars for South Texas border cities of Brownsville and Mission.

Fine particles, $PM_{2.5}$, can be associated with mobile sources. $PM_{2.5}$ clearly shows an increasing trend for both Brownsville and Mission with concentrations

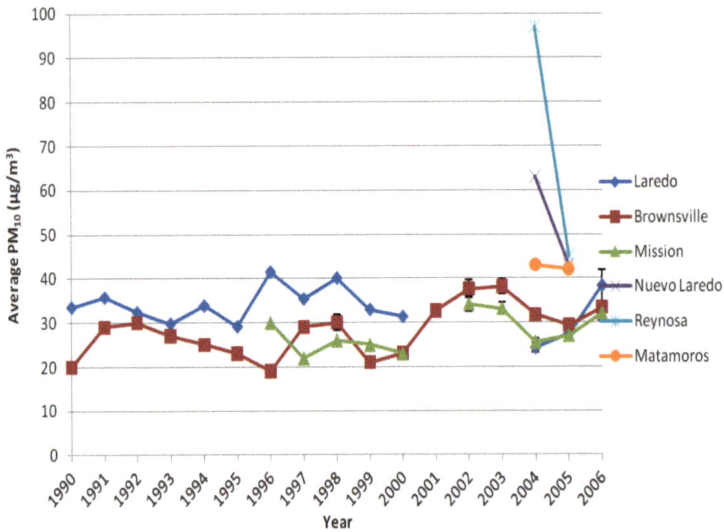

Fig. 3.5 PM_{10} concentrations for South Texas – Northeastern Mexico border cities

trending closer to the NAAQS of 15 μg/m^3. The concentration differences between the two cities were not statistically significant based on the standard errors bars. No $PM_{2.5}$ data was available for Tamaulipas; however, data for coarse particles, PM_{10}, was available for the Tamaulipas cities of Nuevo Laredo and Reynosa for 2004 and 2005. Figure 3.5 shows the comparison of PM_{10} between the border cities of South Texas and Northeastern Mexico.

PM_{10} concentrations for South Texas cities have remained stable for the last 17 years. For the 1 year data was available, it was apparent that the Tamaulipas cities exhibited higher levels of PM_{10} than the border cities of South Texas.

3.3.2 *Health Statistics Data*

3.3.2.1 Mortality Cases and Rates for South Texas – Northeastern Mexico Border Region

The mortality health statistics by type of disease for the South Texas – Northeastern Mexico border region that were considered included cardiovascular and respiratory diseases. The two types of cardiovascular diseases considered in this study were heart disease and stroke. The two types of respiratory diseases' mortality rates considered were lung disease and CLRD. Table 3.3 indicates the mortality cases due to heart disease for South Texas counties (TDSHS, 2007). There was a general increasing trend for mortality cases due to heart disease in South Texas and more cases occurred in the most populated counties of Hidalgo, Cameron, and Webb than in the other smaller counties.

Table 3.3 Mortality cases due to heart disease for 7 South Texas counties (TDSHS, 2007)

Selected counties of South Texas								
Year	Cameron	Hidalgo	Starr	Webb	Willacy	Zapata	Zavala	Total
1995	438	697	61	225	29	28	24	1,502
1996	441	794	61	211	41	26	37	1,611
1997	560	822	62	259	33	37	26	1,799
1998	548	817	73	237	32	45	31	1,783
1999	493	774	69	222	32	35	31	1,656
2000	508	784	74	228	30	35	35	1,694
2001	573	931	100	268	23	35	31	1,961
2002	627	913	95	282	32	30	22	2,001
2003	581	913	92	280	28	22	24	1,940

Table 3.4 Mortality cases due to heart disease for 3 major cities of Tamaulipas (Office of the Secretaria de Salud del Estado de Tamaulipas)

Cities of Tamaulipas				
Year	Nuevo Laredo	Reynosa	Matamoros	Total
2001	10	79	8	97
2002	33	53	45	131
2003	28	143	42	213
2004	34	18	27	79
2005	20	6	14	40
2006	107	90	46	243

Table 3.4 indicates the mortality cases due to heart disease for the major cities of Tamaulipas collected from the Office of the Secretaria de Salud del Estado de Tamaulipas through personnel communication. There was a generally increasing trend in mortality cases due to heart disease for all major Tamaulipas cities from 2001 to 2006.

Figure 3.6 shows an overall comparison between the two sides of the border for mortality rates due to heart disease. The data represents mortality rates per 100,000 people. The mortality rates for heart disease were significantly higher in the 7 South Texas counties than in the 3 Northeastern Mexico border cities based on observation.

Mortality cases due to stroke in the seven South Texas counties and three major cities of Tamaulipas are presented in Tables 3.5 and 3.6, respectively.

Hidalgo and Cameron, which were the most populated counties in the study region, exhibited more mortality cases due to stroke than in the other smaller counties in South Texas. For the reported years, mortality cases due to stroke were higher in Reynosa than in Nuevo Laredo and Matamoros.

Figure 3.7 presents a comparison for mortality rates due to stroke between South Texas and Northeastern Mexico. The mortality rates for stroke were significantly higher in South Texas than in Northeastern Mexico.

Table 3.7 indicates the mortality cases due to lung cancer for South Texas. Mortality cases due to lung cancer for the major cities of Tamaulipas were only available for the year 2001. There were four mortality cases for Nuevo Laredo and nine for Matamoros. No mortality cases were reported for Reynosa. There were

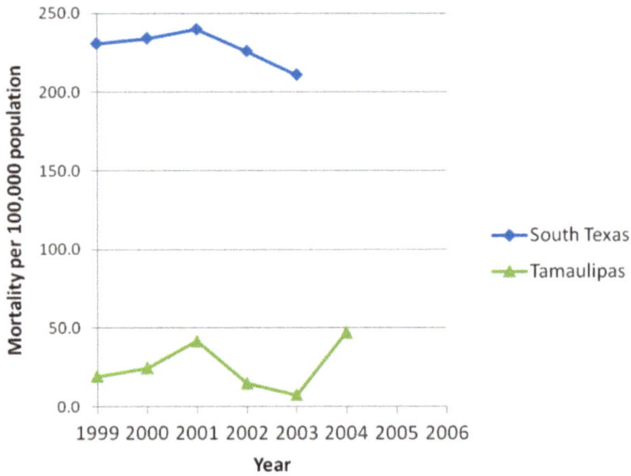

Fig. 3.6 Heart disease mortality rates for 7 South Texas counties – 3 Northeastern Mexico border cities

Table 3.5 Mortality cases due to Stroke for 7 South Texas counties (TDSHS, 2007)

Selected counties of South Texas								
Year	Cameron	Hidalgo	Starr	Webb	Willacy	Zapata	Zavala	Total
1995	120	156	18	55	12	3	10	374
1996	95	163	8	63	8	5	6	348
1997	123	149	9	67	14	4	6	372
1998	127	125	15	58	9	4	6	344
1999	123	174	18	61	9	0	7	392
2000	119	133	16	55	6	5	8	342
2001	123	148	11	73	8	5	5	373
2002	129	175	15	66	4	2	6	397
2003	101	161	11	55	7	4	4	343

Table 3.6 Mortality cases due to stroke for 3 major cities of Tamaulipas (Office of the Secretaria de Salud del Estado de Tamaulipas)

Cities of Tamaulipas				
Year	Nuevo Laredo	Reynosa	Matamoros	Total
2001	4	68	5	77
2002	4	32	–	37
2003	–	71	7	78
2004	–	–	–	0
2005	–	26	–	26
2006	–	91	–	91

more mortality cases due to lung cancer in the most populated counties of Hidalgo and Cameron than in the other smaller border counties.

Figure 3.8 presents the mortality rates due to lung cancer for South Texas and Northeastern Mexico. In 2001, the only year with available data, the mortality rates for lung cancer were significantly higher in South Texas than in Northeastern Mexico.

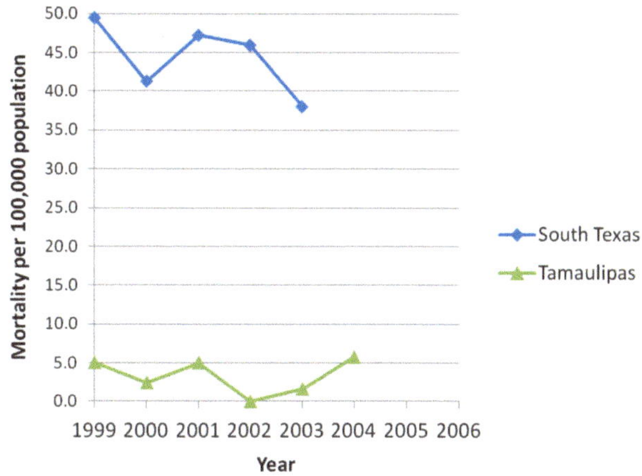

Fig. 3.7 Stroke mortality rates for 7 South Texas counties – 3 Northeastern Mexico border cities

Table 3.7 Mortality cases due to lung cancer for 7 South Texas counties (TDSHS, 2007)

Selected counties of South Texas								
Year	Cameron	Hidalgo	Starr	Webb	Willacy	Zapata	Zavala	Total
1995	80	130	12	21	8	3	0	254
1996	88	97	7	30	2	4	3	231
1997	98	124	10	37	2	4	3	278
1998	96	117	12	37	7	3	3	275
1999	83	124	7	33	13	2	3	265
2000	83	127	9	21	7	2	1	250
2001	89	111	6	28	6	2	3	245
2002	85	109	9	33	6	1	4	247
2003	88	131	11	38	4	5	2	279

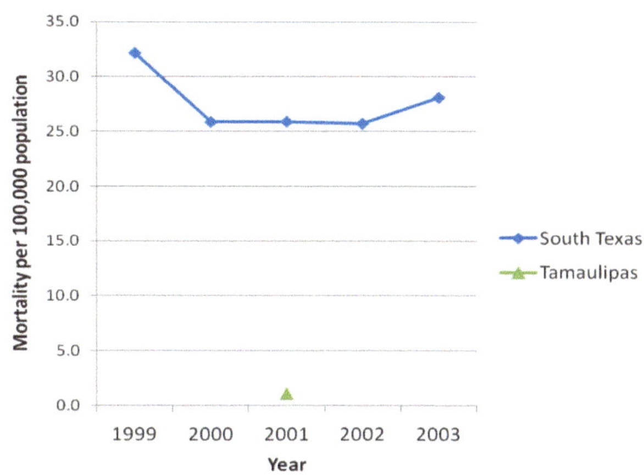

Fig. 3.8 Lung Cancer mortality rates for 7 South Texas counties – 3 Northeastern Mexico border cities

Table 3.8 Mortality cases due to CLRD for 7 South Texas counties (TDSHS, 2007)

Selected counties South Texas								
Year	Cameron	Hidalgo	Starr	Webb	Willacy	Zapata	Zavala	Total
1995	86	123	7	19	5	3	1	244
1996	52	107	12	20	3	1	3	198
1997	63	115	8	9	8	2	4	209
1998	58	113	7	13	5	4	2	202
1999	63	124	9	23	4	6	2	231
2000	57	106	5	21	5	3	1	198
2001	63	112	13	16	1	2	2	209
2002	66	126	12	15	5	1	1	226
2003	67	126	14	26	7	4	3	247

Table 3.9 Mortality cases due to CLRD for 3 major cities of Tamaulipas (Office of the Secretaria de Salud del Estado de Tamaulipas)

Cities of Tamaulipas				
Year	Nuevo Laredo	Reynosa	Matamoros	Total
2001	11	246	23	279
2002	2	103	34	139
2003	9	100	9	118
2004	30	–	54	84
2005	–	–	41	41
2006	60	304	72	436

Tables 3.8 and 3.9 indicate mortality cases due to CLRD for seven South Texas counties and three major cities of Tamaulipas, respectively. The data indicates that mortality cases for CLRD were higher in Hidalgo and Cameron than in the other border counties of South Texas, most likely because those two counties have larger populations.

Mortality cases due to CLRD were higher in Reynosa than in Nuevo Laredo and Matamoros for the years for which data was available. Figure 3.9 shows the mortality rate due to CLRD for South Texas – Northeastern Mexico border region. The mortality rates for CLRD were significantly higher in South Texas than in Northeastern Mexico.

3.3.2.2 Morbidity Cases and Rates Due to Asthma for South Texas – Northeastern Mexico Border Region

The morbidity cases and morbidity rates due to asthma for the 7 South Texas counties and the 3 Northeastern Mexico border cities were considered. Tables 3.10 and 3.11 indicate morbidity cases due to asthma for South Texas and major cities of Tamaulipas, respectively, and Fig. 3.10 shows a comparison of asthma morbidity rates between the two sides of the border region.

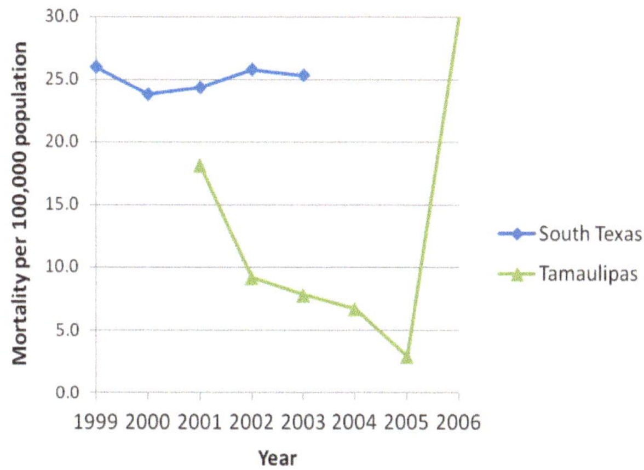

Fig. 3.9 CLRD mortality rates for 7 South Texas counties – 3 Northeastern Mexico border cities

Table 3.10 Morbidity cases due to asthma for South Texas (TDSHS, 2007)

Selected counties of South Texas					
Year	Cameron	Hidalgo	Webb	Willacy	Total
2001	235	244	123	22	624
2002	232	262	109	21	624
2005	271	278	136	22	707

Table 3.11 Morbidity cases due to asthma for major cities of Tamaulipas (Office of the Secretaria de Salud del Estado de Tamaulipas)

Cities of Tamaulipas				
Year	Nuevo Laredo	Reynosa	Matamoros	Total
2004	240	284	196	720
2005	217	617	255	1,089
2006	170	467	280	917

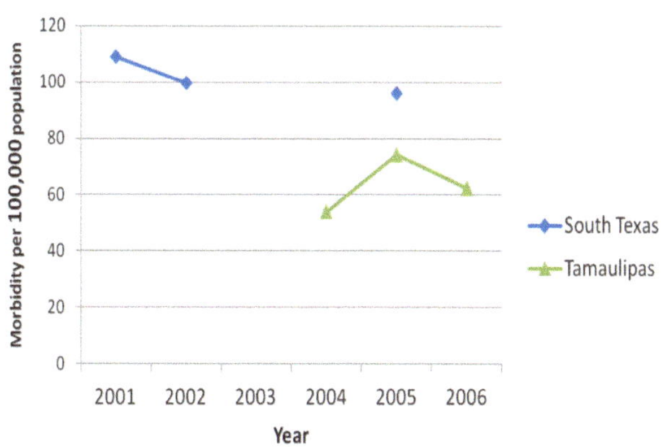

Fig. 3.10 Comparisons of asthma morbidity rates for South Texas and Northeastern Mexico

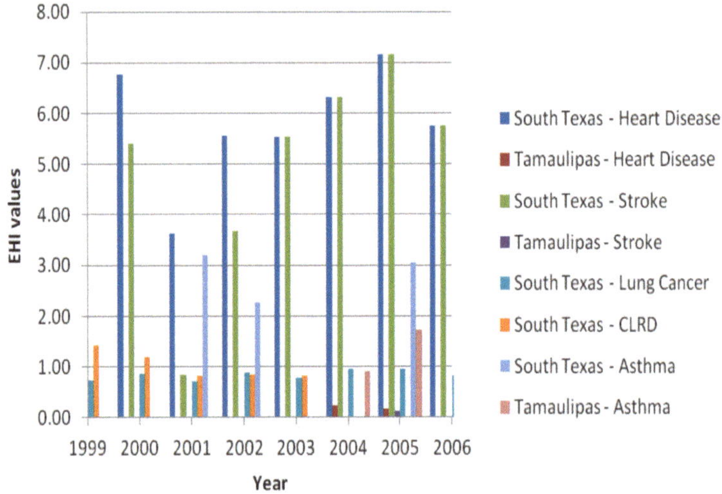

Fig. 3.11 EHI-PM_{10} by disease type

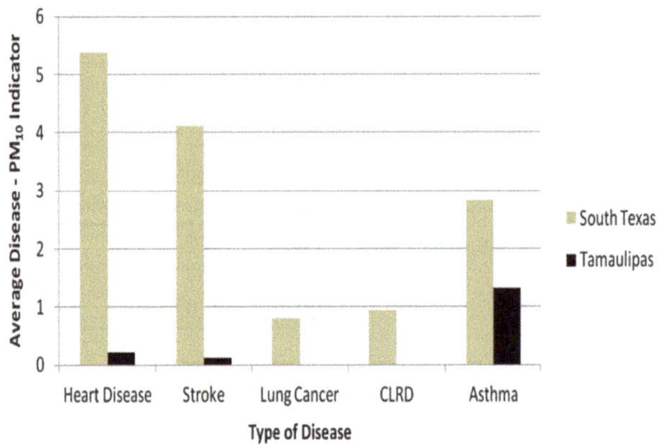

Fig. 3.12 Comparison of EHI-PM_{10} for different diseases (1999–2006 average)

3.3.3 Calculation of Environmental Health Indicators (EHI)

EHI were defined for this study as ratios of mortality rates, due to heart disease, stroke, lung cancer, CLRD and the environmental burden of interest in concentration units (μg/m^3). Figure 3.11 presents time series EHIs for PM_{10} between 1999 and 2006.

Figure 3.11 indicates no clear trends in the EHI-PM_{10} in the time series 1999–2006; however, the indicators for South Texas related to heart disease and stroke seemed relatively high, and the indicator for asthma appeared relatively high both in South Texas and in Tamaulipas. To examine those observations more clearly, the average indicator values are presented in Fig. 3.12. This also allows,

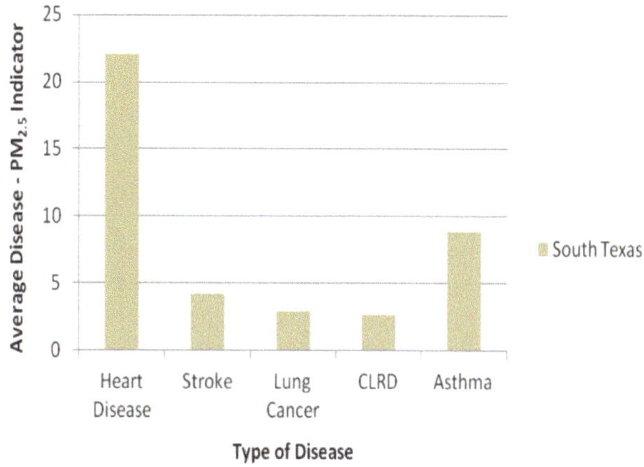

Fig. 3.13 Comparison of EHI-$PM_{2.5}$ values for different diseases in South Texas

Table 3.12 *t*-Test results of correlating air quality and health diseases

Correlation tested	R^2	*t*-Value	*t*-Critical	Null hypothesis result
CO with Lung Cancer	0.5926	2.409	2.132	Rejected
CO with Asthma	0.3897	0.7979	6.314	Not rejected
O_3 with Heart Disease	0.9313	6.3617	2.353	Rejected
O_3 with Stroke	0.3834	1.5755	2.132	Not rejected
$PM_{2.5}$ with CLRD	0.3819	1.1094	2.920	Not rejected
$PM_{2.5}$ with Asthma	0.4164	0.834	6.314	Not rejected

for a comparison of the EHI-PM_{10} between South Texas and Tamaulipas for each type of disease considered in the study.

EHI-PM_{10} values for South Texas are higher for heart disease, stroke, and asthma when they are compared to lung cancer and CLRD. The indicator of asthma for Tamaulipas is significantly higher as compared to other diseases in Tamaulipas. The indices show that asthma was prevalent along the South Texas – Northeastern Mexico border region.

Figure 3.13 presents a similar analysis of the EHI-$PM_{2.5}$ (cases per 10^5 people – $\mu g/m^3$) in South Texas only since no $PM_{2.5}$ data was reported in Tamaulipas.

The EHI values for $PM_{2.5}$ show that the indices for heart disease and asthma were significantly higher when compared to stroke, lung cancer, and CLRD. The indices, for heart disease and asthma were about four times that of PM_{10} as shown in Fig. 3.12.

3.3.4 Correlation of Air Quality Data with Health Statistics

To evaluate the significance of any correlation between air pollutants in the border region and health effects, correlations between air quality data and epidemiologic health statistics were assessed by means of scatter plots. To check the significance

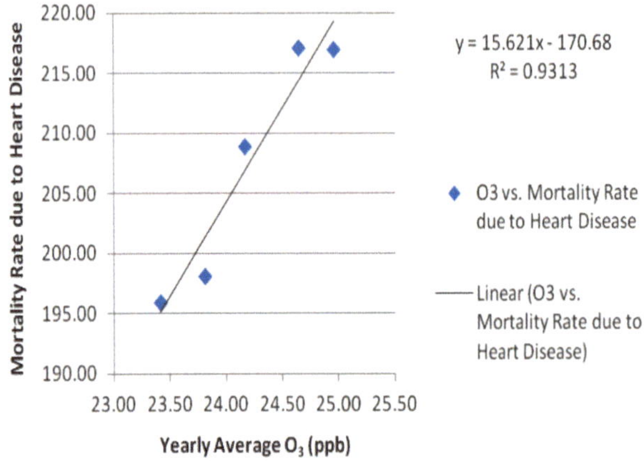

Fig. 3.14 Relationship between ozone (O_3) ppb and mortality rate due to heart disease

of the correlation coefficients, 0.95 confidence level *t*-tests were performed with the null hypothesis stating there is no relationship between the selected air pollutant and the health impact considered. The null hypothesis is rejected if the *t*-value is greater than the *t*-critical value. Table 3.12 summarizes the results of the 2-tailed *t*-tests performed to correlate air quality and health diseases.

The linear regression results in Table 3.12 indicated there was a linear relationship between CO and mortality rate due to lung cancer, with about 60 % of the total variation in the mortality rate due to lung cancer explained by its relationship with CO. Carbon monoxide (CO) is not a precursor of lung cancer; however, CO is an indicator of emissions of products of incomplete combustion that can be carcinogenic. What seemed more evident was the significance of the correlation between O_3 and heart disease that exhibits a *t*-value almost three times the value of *t*-critical. This indicated a strong linear relationship between mortality rate due to heart disease and the yearly average O_3 concentration as shown in Fig. 3.14.

The coefficient of determination (R^2) implies that 93 % of the total variation in the mortality rate due to heart disease could be explained by the linear relationship with the yearly average O_3 concentration. The high value of R^2 in this case clearly indicates that as the yearly average O_3 concentration increased, the mortality rate due to heart disease also increased.

The percentages of total variation in the mortality/morbidity rates for various diseases attributed to different air pollutants that remain unexplained may be due to the lack of sufficient air quality and health statistic data.

3.4 Conclusions

Air quality and health data were collected for the South Texas – Northeastern Mexico border region. Environmental health indicators were developed and used in conjunction with statistical correlation analyses to study the impact of air pollutants on human health in this region.

The environmental health indicators due to particulate matter, both PM_{10} and $PM_{2.5}$, for South Texas were higher for heart disease, stroke, and asthma when compared to those for lung cancer and CLRD. The indicator values due to PM_{10} were higher for asthma than any other diseases in Tamaulipas. In general, the study showed that asthma was prevalent along the South Texas – Northeastern Mexico border region during the study period.

The study demonstrated that for the border region, mortality due to heart disease and O_3 concentration exhibited a strong linear correlation that was statistically significant. It also indicated that the mortality rates for heart disease, stroke, lung cancer, and CLRD were significantly higher in South Texas than in Northeastern Mexico. However, this may have been due to insufficient air quality and health statistic data for Tamaulipas. The health data reported by the Office of Secretaria de Salud del Estado de Tamaulipas covered only 39.1 % of the population.

In Northeastern Mexico, the city of Reynosa exhibited the highest mortality due to CLRD than any of the other cities of Tamaulipas. Also, the asthma morbidity rate in Tamaulipas was high and quite comparable to the rates north of the border.

Based on the environmental data available, it was found that the selected South Texas counties met the NAAQS standards for ambient CO and O_3 concentrations; however, $PM_{2.5}$ showed an increasing trend in all South Texas border counties approaching the maximum NAAQS standard. $PM_{2.5}$ data was not available for Tamaulipas, but PM_{10} concentrations were higher in the major cities of Tamaulipas when compared to South Texas.

The proposed environmental health indicators may be used to track and evaluate trends and to support decision makers and researchers. The indicators will benefit from the enhancement of the comparability of the epidemiological information, the air quality data from both sides of the border, and the avoidance of data gaps.

Acknowledgements This material is based upon work supported by the Center of Research Excellence in Science and Technology – Research on Environmental Sustainability of Semi-Arid Coastal Areas (CREST-RESSACA) at Texas A&M University– Kingsville (TAMUK) through a Cooperative Agreement (No. HRD-0206259) from the National Science Foundation (NSF). Any opinions, findings, and conclusions or recommendations expressed in this material are those of the author(s) and do not necessarily reflect the views of the National Science Foundation.

References

Afroz, R., Hassan, M. N., & Ibrahim, N. A. (2003). Review of air pollution and health impacts in Malaysia. *Environmental Research, 92*(2), 71–77.

Beamish, L. A., Osornio-Vargas, A. R., & Wine, E. (2011). Air pollution: An environmental factor contributing to intestinal disease. *Journal of Crohn's & Colitis, 5*(4), 279–286.

Bell, M. L., Cifuentes, L. A., Davis, D. L., Cushing, E., Gusman Telles, A., & Gouveia, N. (2011). Environmental health indicators and a case study of air pollution in Latin American cities. *Environmental Research, 111*(1), 57–66.

Briggs, D., Corvalan, C., & Nurminen, M. (1996). *Linkage methods for environment and health analysis*. Geneva, Switzerland: UNEP/US EPA/WHO.

Hall, J. V., Brajer, V., & Lurmann, F. W. (2010). Air pollution, health and economic benefits – Lessons from 20 years of analysis. *Ecological Economics, 69*(12), 2590–2597.
Kjellstrom, T., & Corvalan, C. (1995). Framework for the development of environmental health indicators. *World Health Statistics Quarterly, 48*(2), 144–154.
Laumbach, R. J., & Kipen, H. M. (2012). Respiratory health effects of air pollution: Update on biomass smoke and traffic pollution. *The Journal of Allergy and Clinical Immunology, 129*(1), 3–11.
Lee, Y. L., Wang, W.-H., Lu, C.-W., Lin, Y.-H., & Hwang, B.-F. (2011). Effects of ambient air pollution on pulmonary function among schoolchildren. *International Journal of Hygiene and Environmental Health, 214*(5), 369–375.
Norman, L. M., Feller, M., & Phillip Guertin, D. (2009). Forecasting urban growth across the United States–Mexico border. *Computers, Environment and Urban Systems, 33*(2), 150–159.
Raysoni, A. U., Sarnat, J. A., Sarnat, S. E., Garcia, J. H., Holguin, F., Luèvano, S. F., et al. (2011). Binational school-based monitoring of traffic-related air pollutants in El Paso, Texas (USA) and Ciudad Juárez, Chihuahua (México). *Environmental Pollution, 159*(10), 2476–2486.
Shi, C., Fernando, H. J. S., & Yang, J. (2009). Contributors to ozone episodes in three U.S./Mexico border twin cities. *Science of the Total Environment, 407*, 5128–5138.
Texas Commission on Environmental Quality [TCEQ]. (2007). *Air monitoring sites*. Retrieved November 15, 2007, from http://www.tceq.com/compliance/monitoring/air/monops/sites/monitors_map.html
Texas Department of State Health Services [TDSHS]. (2007). *Texas health facts* (*County fact sheet*). Retrieved December 1, 2007, from http://www.dshs.state.tx.us/chs/cfs/
United States Environmental Protection Agency [U.S. EPA]. (2006). *State of the border region 2012: U.S.-Mexico environmental program indicators report 2005*. Retrieved May 12, 2010, from http://www.epa.gov/usmexicoborder/indicators/05report/SBR_2005_English.pdf
United States Environmental Protection Agency [U.S. EPA]. (2011). *Border 2012: U.S.–Mexico environmental program*. State of the border region indicators report 2010. Retrieved December 12, 2012, from http://www2.epa.gov/sites/production/files/documents/border-2012_indicator-rpt_eng.pdf
Yang, C., Yang, H., Guo, S., Wang, Z., Xu, X., Duan, X., et al. (2012). Alternative ozone metrics and daily mortality in Suzhou: The China Air Pollution and Health Effects Study (CAPES). *Science of the Total Environment, 426*, 83–89.

Chapter 4
Characterization of Ultrafine Particles and Other Vehicular Pollutants Near an International Bridge in United States – Mexico Border Region

Yungang Wang, Jayanth Kumar Pudota, and Yifang Zhu

Abstract An intensive continuous hotspot study was conducted in the City of Laredo to characterize levels of ultrafine particles (UFP), black carbon (BC), nitrogen oxides (NO_x), ozone (O_3) and carbon monoxide (CO) from December 2007 to January 2008. Data were categorized into three different time periods: (1) Before Christmas, (2) During Christmas, and (3) After Christmas. The mean UFP number concentration throughout the study was 1.87×10^4 per cm^3. Average traffic density was 1,099 per hour. The average concentrations of BC, CO, CO_2, NO_x, and O_3 were 1.825 $\mu g/m^3$, 0.05 parts per million (ppm), 499 ppm, 28 parts per billion (ppb), and 21.6 ppb, respectively. There was a significant correlation between both NO_x and BC with UFP particle count, with $R^2 = 0.63$ and 0.60, respectively. Wind direction was found to have a significant effect on UFP and other pollutants. An exponential decay in UFP (20–290 nm) concentrations with increases in wind speed was observed. Significant correlation ($R^2 = 0.99$) was observed between the vehicle speed and the particle number emission factors. About a 1.5-fold increase in concentration of particle number emission factors was observed when the vehicle speed increased from 16 to 50 km/h.

Y. Wang
Environmental Energy Technologies Division, Lawrence Berkeley National Lab, One Cyclotron Road, MS 90-3058, Berkeley, CA 94720, USA

J. Kumar Pudota
Department of Environmental Engineering, Frank H. Dotterweich College of Engineering, Texas A&M University-Kingsville, 700 University Boulevard, MSC 213, Kingsville, TX 78363, USA

Y. Zhu (✉)
Environmental Health Sciences Department, Jonathan and Karin School of Public School, University of California, Los Angeles, Los Angeles, CA, USA

Center for Clean Air in the Institute of Environment and Sustainability, University of California, Los Angeles, Los Angeles, CA 90095, USA
e-mail: yifang@ucla.edu

D. Ramirez et al. (eds.), *Environmental Sustainability Issues in the South Texas–Mexico Border Region*, DOI 10.1007/978-94-007-7122-2_4,

Keywords Ultrafine particles • Particle number concentration • Particle number emission factor • Vehicular pollutants • Urban air quality

4.1 Introduction

Previous studies have shown associations between particulate matter (PM) exposure and adverse health effects. Particulate air pollution is associated with respiratory and cardiovascular morbidity and mortality (e.g., Polichetti, Cocco, Spinali, Trimarco, & Nunziata, 2009). Increased PM concentrations are associated with increased numbers of cardiovascular related admissions in cities with a great number of PM sources (Bai, Khazaei, Van Eeden, & Laher, 2007). Although it is unlikely that all the physically and chemically diverse forms of inhalable ambient PM have similar toxicity, it is not clear what specific size ranges, chemical species, and/or sources of PM are primarily responsible for health effects. One likely candidate is the ultrafine particle (UFP) defined to be $D_p < 0.1$ μm that is emitted mainly from motor vehicles. When compared with larger particles, UFPs have greater number concentration, surface area and concentrations of adsorbed or condensed toxic air pollutants (oxidant gases, organic compounds, and transition metals) per unit mass (Sioutas, Delfino, & Singh, 2005). Some studies showed that combustion of fossil fuel by vehicles is the primary contributor to UFP in urban environments (Palmgren, Wåhlin, Kildeso, Afshari, & Fogh, 2003). Both diesel and gasoline engines generate a large number of particles in the ultrafine size range (Kittelson, Watts, & Johnson, 2004).

Worldwide, many studies have been conducted to characterize traffic-related UFP in street canyons; at rural, suburban, urban, and urban background sites; and near roadways (e.g., Zhu, Hinds, Kim, & Sioutas, 2002). Most of these studies reported hourly average concentrations from 20,000 to 220,000 per cm^3 at street canyons and from 9,000 to 20,000 per cm^3 in urban backgrounds.

In the Netherlands, Weijers, Khlystov, Kos, and Erisman (2004) found that the UFP concentration increased with increasing traffic density. In Helsinki, Finland, Kerminen et al. (2007) found that the traffic emission was a dominant source of UFP under suitable meteorological conditions and could affect submicron particle number concentrations over vast areas around major roads. In Canada, Diamond and Parker (2004) monitored air quality in the Windsor area from fall 2002 to summer 2003 where heightened security at the ambassador bridge international border in Windsor resulted in long delays especially for the heavy-duty diesel trucks that use this point of entry into the United States. This Canadian study found that during normal traffic movement (no delays) the average increase in total PM near roadway was minimal and the extent of the increased total PM was dependent on traffic density, length of delays, and meteorological conditions (Diamond & Parker).

Continuous measurements of UFP size distributions and other gas phase pollutants including carbon monoxide (CO), nitric oxide (NO) and nitrogen dioxide (NO_2) were conducted in El Paso, Texas on the United States (U.S.)–Mexico border in winter 1999. Throughout the study, the mean UFP number concentration was

1.44×10^4 per cm^3. Significant correlations between CO and UFP ($r = 0.81$) and between NO and UFP ($r = 0.74$) were observed. Most pollutants were found to vary on diurnal cycles and follow the trend of either vehicular traffic densities or sunlight intensity. Wind direction was found to have an influence on pollutant concentrations (Noble et al., 2003).

Despite these near roadway and border-crossing studies, the relationships between traffic densities, meteorological parameters, spatial and temporal variability, and UFP number size distribution near roadways in the border regions are still not well understood. The U.S.–Mexico border area has experienced a surge in population and traffic volume in the past several decades. The local residents and border travelers are exposed to high levels of vehicular air pollutants. Therefore, more measurements of vehicular air pollutants in the U.S.–Mexico border region are needed. The current study comprehensively and systematically assessed the vehicular UFP level and other vehicular pollutants near an international bridge in Laredo to improve the understanding of sources in this region during the Christmas holiday season when the maximum variation of traffic density was observed and provided valuable data to investigate urban UFP profiles. Additional and improved statistical information on the local-scale variability of UFP will improve our knowledge of human exposure to and the health impacts of UFP. It will enable us to make effective control strategies and establish emission standards that will protect public health.

4.2 Experimental Techniques and Methods

4.2.1 Description of the Sampling Site

Laredo is the county seat of Webb County, Texas, United States of America (U.S.A.) on the north bank of the Rio Grande. According to the 2010 U.S. Census (United States Census Bureau, 2000), the city population was 236,091 making it the tenth most populous city in the state of Texas and 3rd most populated on the United States-Mexican border, after San Diego, California and El Paso, Texas. The city has a total area of 206.0 km^2. Laredo is the largest inland port on the U.S.–Mexico border with four International Bridges and one Railway Bridge. The field campaign was performed in the City of Laredo from December 23, 2007, to January 2, 2008. Measurements were conducted near International Bridge No. 2 in Laredo (Fig. 4.1). The average temperature was 13.1 °C, relative humidity (RH) was 40 %, and prevailing wind direction was southeasterly with an average wind speed of 2.2 m/s.

Main reasons for choosing this Laredo site included its close proximity to the International Bridge, the Interstate Highway (IH) 35, and the variation of traffic density in this area during the study period. The International Bridge No. 2 is one of the most important inland ports between the U.S. and Mexico with a monthly average non-commercial traffic flow of 74,538 (Texas Department of Transportation [TX DOT], 2006). Variation of the traffic density during the

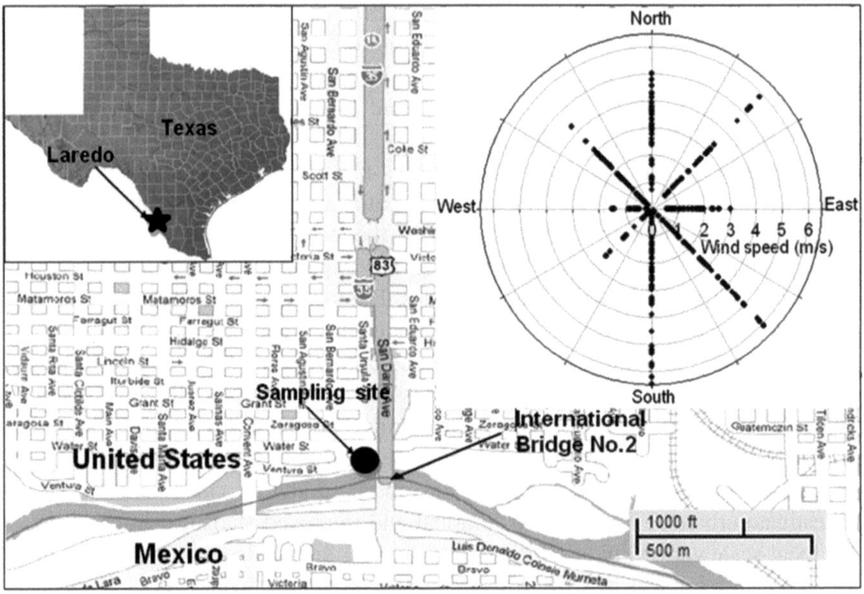

Fig. 4.1 Map of the sampling site, International Bridge No. 2, and the wind direction patterns in Laredo, TX

sampling period was great due to the Christmas holidays. It provided data for a wider traffic range allowing a better investigation of the impact of different traffic densities on UFP concentrations.

4.2.2 Sampling Procedures and Instrumentation

Particle number size distributions were measured by two scanning mobility particle sizers – (a) SMPS 3936 with DMA 3080N and CPC 3785 (TSI Inc.) and (b) SMPS 3936 with DMA 3080L and CPC 3785 (TSI Inc.). Data analysis of the SMPS output were done by the Aerosol Instrument Manager software (version 7.3.0.0, TSI Inc.). SMPS samples were collected using a 100-s up-scan and 20-s retrace. A portable Aethalometer (AE-42 tape, Magee Scientific) measures black carbon, BC, mass concentrations every 1 min. CO, CO_2, ambient temperature and relative humidity were recorded with handheld air quality monitors (Q-Trak, TSI Inc.). Nitrogen oxide (NO_x) and ozone (O_3) concentrations were measured by a NO_x analyzer (Model 42C, Thermo Environmental Instruments, Inc.) and an ozone analyzer (API-400E, Teledyne Instruments Inc.), respectively.

Meteorological data, including ambient temperature, relatively humidity, wind speed, and wind direction were also measured at the sampling site (operated by Texas Commission on Environmental Quality (TCEQ), EPA site number: 48-479-0016), located 3 m above ground level. During each measurement, the traffic density on the international bridge, defined as number of vehicles passing per hour, was continuously recorded. Three-minute traffic count samples were manually collected in 15-min intervals. The mean of the four samples is represented as hourly average data. Passenger vehicles, light-duty trucks, medium-duty trucks, and heavy-duty diesel trucks were counted separately to estimate the traffic density by different type of vehicles. SUVs, vans, and 4-tire trucks are counted as light-duty trucks. Vehicles with two axles and six tires (i.e., four tires on the rear axle) were counted as medium-duty trucks. Heavy-duty trucks (i.e., those with three axels or more) were expected to have diesel engines.

4.2.3 Data Analysis

All statistical analyses (correlation and multiple comparisons) were performed using Microsoft Office Excel 2003, Sigma Plot 10.0 (Systat Software Inc.), and Statistic Analysis System 9.0 (SAS Institute Inc). All the pollutants data were categorized in terms of time period. The pollutants and traffic density data for the Laredo study were separated into three groups: before Christmas (BC), during Christmas (DC), and after Christmas (AC). The averages and standard deviations for the UFP number concentrations, size distributions (in three different modes), and other pollutants were computed for each group. The correlations between UFP, other vehicular air pollutants, and traffic density were examined. The UFP number size distributions measured at the sampling site were analyzed using a linear regression model to determine the pattern and variation of concentration with respect to time. The respective correlations between traffic density, wind direction, wind speed, and UFP concentration and size distribution were examined using linear regression to determine their impacts on UFP.

4.3 Results and Discussion

The descriptive statistics of the measured pollutants from the sampling site have been summarized in Table 4.1. During the measurement period, the traffic densities in the sampling area substantially changed.

4.3.1 Traffic Effect on UFP and Other Vehicular Pollutants

Measured traffic densities are shown in Fig. 4.2. The three curves show a similar trend with the traffic peak hours occurring from 13:00 to 17:00. The daily average traffic densities were given in Table 4.1. Passenger car and light-duty truck traffic densities measured at the site were correlated ($R^2 = 0.57$). These two vehicle types together accounted for 99.5 % of the total traffic volume. The medium-duty trucks

Table 4.1 Average traffic densities, number concentrations of particles (unit: 10^4 per cm^3) and concentrations of other vehicular pollutants (24 h) with one standard deviation in parenthesis

Pollutants	All days	BC	DC	AC
Traffic density (h^{-1})	1,099 (658)	1,376 (783)	899 (424)	1,178 (670)
Nuclei mode (7–20 nm)	0.37 (0.36)	0.59 (0.56)	0.21 (0.16)	0.25 (0.20)
Aitken mode (20–100 nm)	1.26 (0.66)	1.98 (0.81)	0.91 (0.69)	0.10 (0.20)
Accumulation mode (100–290 nm)	0.26 (0.53)	0.50 (0.77)	0.31 (0.41)	0.14 (0.14)
Total UFPs (7–290 nm)	1.89 (1.13)	3.06 (1.23)	1.42 (1.07)	1.42 (1.02)
BC ($\mu g/m^3$)	1.825 (2.200)	2.829 (4.061)	2.695 (2.253)	1.890 (1.872)
CO (ppm)	0.06 (0.12)	0.07 (0.12)	0.02 (0.04)	0.04 (0.07)
CO_2 (ppm)	499 (40)	478 (27)	524 (18)	497 (20)
NO_x (ppb)	28 (15)	45 (29)	29 (10)	18 (10)
Ozone (ppb)	21.6 (12)	20.5 (11)	21.9 (14)	21.5 (11)

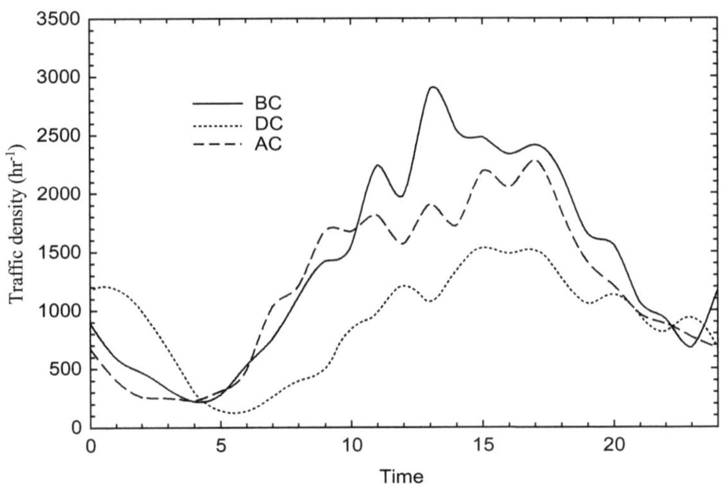

Fig. 4.2 Average hourly traffic on the International Bridge No. 2 during three different sampling periods, before Christmas (*BC*), during Christmas (*DC*), and after Christmas (*AC*)

and heavy-duty diesel trucks occupied the remaining 0.5 %. Light-duty trucks alone accounted for 64 % of the total traffic volume.

The measured hourly average UFP number concentrations in 7–20 nm, 20–100 nm, and 100–290 nm size bins and traffic densities observed before Christmas, during Christmas and after Christmas were included in Table 4.1. The UFPs were divided into three modes based on the particle diameter, 7–20 nm (nuclei mode), 20–100 nm (Aitken mode), and 100–290 nm (accumulation mode). It can be seen that the overall particle number concentration of the Aitken mode (20–100 nm) was much higher than for the other two modes. It was noted that particle number concentrations for all the three modes were highest before Christmas when the traffic density was highest. The lowest particle number concentrations for the three modes were observed during Christmas when the traffic

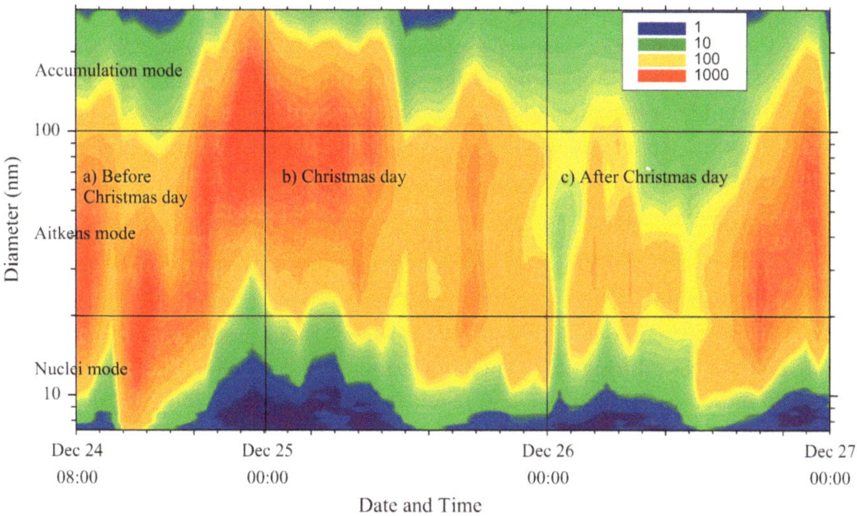

Fig. 4.3 Mean diurnal variation of UFP size distribution with diameter on the y-axis, time of day on the x-axis, and particle concentration by shading contour for (**a**) before Christmas, (**b**) during Christmas day, and (**c**) after Christmas

density was relatively low. Particles in the size range of 10–30 nm are emitted in large numbers by diesel engines (Kittelson, 1998). A strong correlation between the number concentrations of UFPs and traffic volume at a busy urban intersection has been reported (Wang et al., 2008). The strong association between particle concentrations in the Aitken mode and traffic density may support the hypothesis that vehicular emissions were the dominant source of UFP for this sampling area. Table 4.1 also summarizes the average concentrations of other vehicular air pollutants.

4.3.2 Diurnal Variation of UFP

Previous studies suggested that diurnal patterns of UFP number concentrations are influenced predominately by two main factors: (1) meteorological conditions, and (2) local sources such as traffic, coal-fire power plant, and related particle nucleation (Morawska, Jayaratne, Mengersen, Jamriska, & Thomas, 2003; Wang, Hopke, Chalupa, & Utell, 2011). Figure 4.3 displays the diurnal patterns of UFP number concentrations with the time of the day on the *x*-axis, the particle size on the *y*-axis (logarithmic scale). The color intensity indicates normalized particle number concentration (dN/dLogDp) from December 24–27, 2007. Due to an instrument problem, no data were collected between 00:00 and 08:00 h on December 24, 2007. As shown in Fig. 4.3, particles in Aitken mode showed a high number concentration in the late afternoon (18:00–19:00) for all the 3 days.

From field observations, the nearby bridge was filled with cars and trucks between 1800 and 1900. Many people were sitting in their vehicles waiting to pass the U.S. custom. The whole bridge turned into a large parking lot. However, this event could not be reflected from our traffic record, since there were less traffic passing through compared to other periods. It has been well documented that idling vehicles can emit greater amount of UFPs compared to those with constant cruising speed (e.g., Wang et al., 2008). Therefore, the peak in the UFP during 1800 to 1900 was mainly due to this traffic change. Strong photochemical activity may have favored secondary particle formation and growth in the late afternoon (Gao et al., 2007).

The diurnal variation of the particles in the Aitken mode was found to be a multi-peak pattern during the sampling period (07:00–24:00). The concentration began to increase after sunrise and reached the first peak at about 09:00 and the second peak around 19:00. Both of the two peaks appeared during rush hour traffic. The shoulder peak appeared at early afternoon hours (14:00) with an increase in nuclei mode possibly caused by the growth of UFP under strong photochemical promotion. A similar trend was observed at 18:00 on both the nuclei mode and Aitken mode curves. The pre-existing larger particles depress the formation of new particles due to their coagulation effect on surface which may explain the decrease of UFP after 14:00 (Gao et al., 2007). This may be a possible reason for the higher number concentration for particles in the Aitken and accumulation modes from 0:00 to 8:00. Before Christmas, the UFP concentration in the diameter range of 20–40 nm reached the peak values at morning and noon. During Christmas (a low traffic density day), the peak value for particles in the same size range showed up in the late afternoon. The highest particle concentration throughout the day was around 100 nm and occurred at midnight. The biomass burning and/or the fireworks on Mexico side may have been the possible sources for this peak concentration (southwesterly wind). The particle number concentrations after Christmas were much lower for all the sizes than during the first 2 days. The peak concentrations occurred in the late afternoon in the Aitken mode.

4.3.3 Diurnal Variation of Other Vehicular Pollutants

Besides UFP number size distribution, the concentrations of black carbon (BC), nitrogen oxides (NO_x), carbon monoxide (CO), and ozone (O_3) were also measured.

4.3.3.1 Black Carbon

BC is one of the main products among the pollutants emitted from motor vehicles, and it has been used as an indicator of exposure to diesel exhaust (Bhugwant, Cachier, Bessafi, & Leveau, 2000; Zhu et al., 2002). Mean BC concentrations for

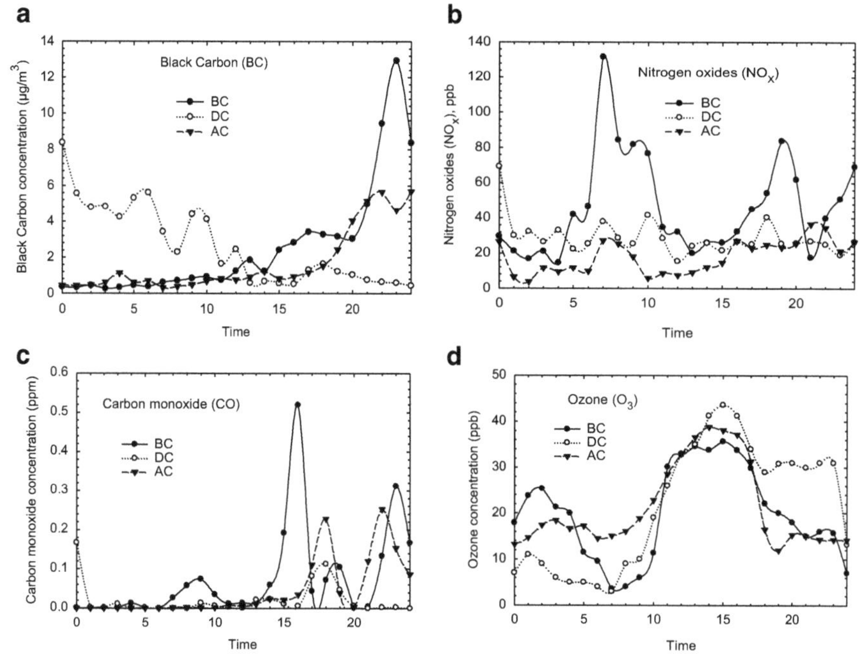

Fig. 4.4 Average hourly concentrations of (**a**) BC, (**b**) NO_x, (**c**) CO, and (**d**) O_3 at sampling site during three different sampling periods

different periods are shown in Fig. 4.4a. Concentrations found from 20:00 on December 24, 2007 to 12:00 on December 25, 2007 were the highest among the three periods. The second highest concentration was observed from 19:00 to 23:00 during the period of after Christmas Day. For the "During Christmas" and "After Christmas" periods, the average traffic density decreased by a factor of about 1/3 to 1/2 compared with the "Before Christmas" period (Fig. 4.2). There could be some other sources contributing to the BC concentration increase (e.g. biomass and fireworks) in this region.

4.3.3.2 Nitrogen Oxides

Mean concentrations of NO_x for three different periods at sampling site are shown in Fig. 4.4b. There were two peaks on the curve for before Christmas Day, one occurred at 7:00 and the other at 17:00. The temperature inversion layer in the early morning and high traffic density in the late afternoon may contribute to these two peak concentrations, respectively. For the remaining two periods, the overall average concentrations were much lower than the one before Christmas and showed a similar trend.

Table 4.2 Correlation (R^2) matrix for UFP and other vehicular air pollutants

	BC	CO	NO	NO_x
UFP	0.60	0.39	0.58	0.63
BC	–	0.36	0.12	0.11
CO	–	–	0.22	0.27
NO	–	–	–	0.95

4.3.3.3 Carbon Monoxide

CO, as a relatively inert gas, has been used as a tracer for traffic air pollution (Chan, Chan, & Qin, 1999). Figure 4.4c displays the average hourly CO concentrations during three different periods. There was one clear peak at 16:00 before Christmas. For the remaining two periods, the peak concentrations occurred at 19:00 which corresponded to the CO concentration measured by the TCEQ. Nearby biomass combustion activities could be the cause given the absence of large traffic emissions during that time frame. The morning traffic on December 24, 2007 also had impact on the CO concentration.

4.3.3.4 Ozone

Concentration profile of ozone is shown in Fig. 4.4d. The ozone levels tend to follow the solar radiation intensity, resulting in higher ozone concentrations during the daylight period. In these cycles, the increase in ozone levels during daylight was due to the combined effects of photochemical production of ozone in the mixing layer and the transport from upper layers (Zunckel et al., 2004), which is favored at noon by the convective activity in the continental boundary layer. Both mechanisms are activated by solar radiation. The lower nocturnal ozone levels were attributed to in-situ destruction of ozone by the well-known reaction between O_3 and NO (Pratt et al., 1983). The ozone hourly average concentrations in Laredo reached maximum values mainly during late afternoon (Fig. 4.4d).

4.3.4 Correlation Analysis Between UFP and Other Vehicular Pollutants

4.3.4.1 Correlation Analysis

The coefficients of determination between UFP and other vehicular pollutants are shown in Table 4.2. UFP showed little dependency on CO but was dependent on BC, NO, and NO_x. The highest correlation coefficient value was observed between UFP and NO_x. NO_x is a co-pollutant from vehicular emission. Morawska et al. (2003) found that both NO_x and UFP were correlated with traffic density on the freeway 100 m from the sampling site when the freeway was upwind. Ketzel,

Table 4.3 Results of cluster analysis of wind direction

					Aerosol (10^4 per cm^3)		
Cluster	WD	WS (m/s)	Temp (°C)	NO_x (ppb)	7–20 nm	20–100 nm	100–290 nm
1	19	2.7	11.8	38	0.5285	1.1669	0.0676
2	71	1.8	12.3	32	0.3782	1.18927	0.2073
3	124	1.8	11.4	29	0.1355	1.6871	0.5343
4	157	3.9	17.2	25	0.2269	0.8316	0.2505
5	198	1.3	11.0	21	0.0893	1.0988	0.4710
6	254	1.1	10.7	36	0.2558	1.2932	0.3284
7	291	0.9	10.1	41	0.2928	1.9613	0.9722
8	339	2.9	16.1	26	0.3543	1.0071	0.1478

Wahlin, Berkowicz, and Palmgren (2003) reported that NO_x was highly correlated ($R^2 = 0.86$) with total particle concentration (less than 200 nm) at street level in Copenhagen. Therefore, it may be possible to predict UFP concentrations from NO_x concentrations on an hourly basis.

4.3.4.2 Using Nitrogen Oxides Data to Indicate UFP Number Concentrations

Table 4.3 shows the results of cluster analyses. Eight clusters were identified for robust statistical analysis of NO_x and UFP concentration variation as a function of wind direction. Cluster 4 and 5 provide the lowest mean NO_x and particle number concentrations (20–100 nm) when the sampling site became the downwind side of the International Bridge No. 2. Cluster 7 provides the highest mean NO_x and particle concentration (20–100 nm) when the wind was blowing from the northwest.

Figure 4.5 shows scatter plots of particle counts against NO_x concentrations. Large differences in the gradient of the linear regression among the different particle size ranges were observed. Low correlation ($R^2 = 0.07$) was found for the accumulation mode (100–290 nm). The correlation for the Aitken mode (20–100 nm) was $R^2 = 0.33$, and the nuclei mode (7–20 nm) produced the strongest correlation ($R^2 = 0.36$). The data collected when the wind speed was above 2.0 m/s and the ambient temperature was above 13 °C were removed to minimize the meteorological effects, resulting in substantial improvements in correlations for both the nuclei mode ($R^2 = 0.89$) and the Aitken mode ($R^2 = 0.45$). In this part of the study, only 7–100 nm particles were considered, a size range dominated by vehicle-emitted particles.

Figure 4.5b shows the scatter plot and linear regression for the hourly average particle concentration (7–100 nm) with the NO_x concentration. The correlation ($R^2 = 0.75$) is much stronger than all the R^2 values in Fig. 4.5a. This indicates the importance of carefully defining the particle size range to derive a credible relationship with NO_x concentration.

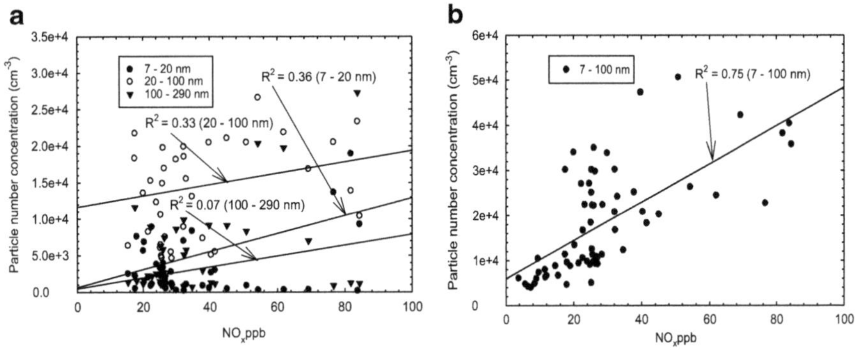

Fig. 4.5 Scatter plots of clustered arithmetic means of (**a**) particle concentrations against NO_x concentrations and (**b**) particle concentrations in the size range of 7–100 nm against NO_x concentrations

4.3.5 *Effects of Wind on UFP and Other Vehicular Air Pollutants*

4.3.5.1 Wind Direction Effect

Figure 4.6a shows the possible locations of sources contributing to the 7–100 nm particles measured at the sampling site. The pollutant data were assigned into eight different groups based on wind directions. The highest particle number concentrations were observed for all modes when the wind was northwesterly. Downtown Laredo was situated to the northwest of the sampling site. The IH-35 and other local roadways were located to the north of the sampling site. Therefore, the emissions from Laredo traffic are believed to have more impact on the UFP concentrations than the border crossing traffic. Dependences of other vehicular pollutants on wind direction were shown in Fig. 4.6b–d. The BC concentration was more dependent on the southwesterly and southeasterly wind directions. It is necessary to note that the International Bridge No. 1 and the railway lines were located to the southwesterly side of the sampling site, which can be also be important sources for BC in addition to Laredo downtown traffic, the biomass burning, and fireworks across the border in Mexico. High concentrations for both NO_X and CO were observed from the north that could be due to downtown traffic and IH-35 effects.

4.3.5.2 Wind Speed Effect

Previous studies have suggested that wind speed was an important factor determining the concentration of UFP (Wang et al., 2008; Zhu et al., 2002). Figure 4.7 shows the particle concentrations as a function of the wind speed for the particles in four different size bins. Except for the particles in nuclei mode in Fig. 4.7a, the particle concentrations in the other three modes exhibited a clear exponential decay trend.

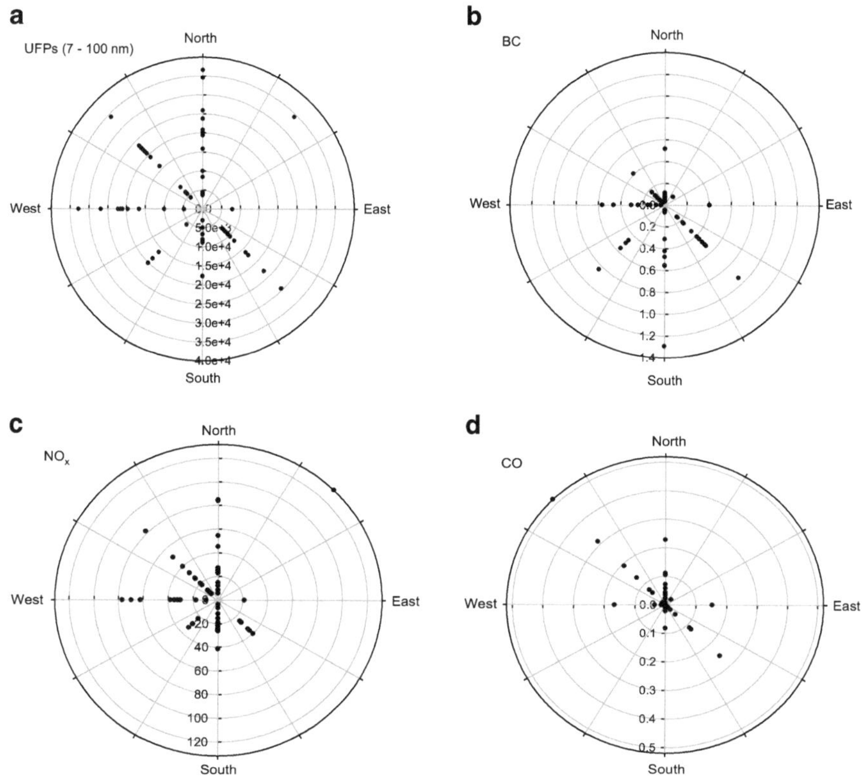

Fig. 4.6 Source estimation of various air pollutants based on wind direction at sampling site: (**a**) UFPs (7–100 nm/cm^3), (**b**) BC (μg/m^3), (**c**) NO_x (ppb), and (**d**) CO (ppm)

4.3.6 Emission Factor of UFP and Comparison with Other Studies

Wind directions affect UFP number concentrations (Wang et al., 2008). When the wind blew from the southeast, the sampling site was mainly influenced by the vehicle traveling on the bridge No. 2. On December 27, 2007 the wind was blowing constantly from the southeast (150° ± 11° clockwise from North) from 9:00 to 22:00, and the traffic density was at the average level (1,679 vehicles/h). Therefore, we selected the period from 9:00 to 22:00 on December 27, 2007 to study the particle emission factors from the vehicles travelling on the International Bridge 2. We employed UFP, CO, and CO_2 concentrations measured in the proximity of the sources to derive emission factors for UFP. Assuming that the carbon mass in the vehicles' exhaust is in the form of CO and CO_2, the emission factor can be written as Eq. (4.1) (Ban-Weiss, Lunden, Kirchstetter, & Harley, 2010).

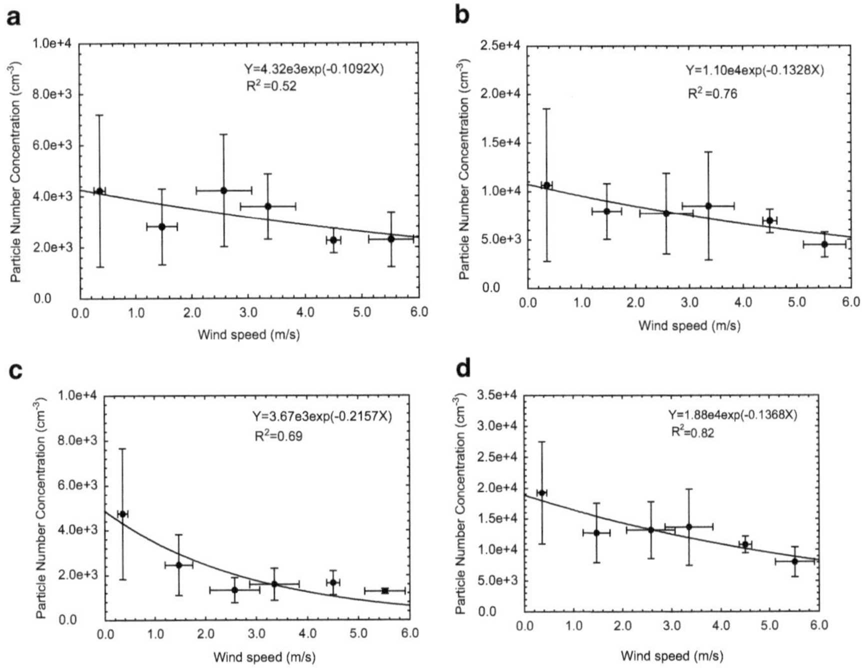

Fig. 4.7 Particle concentrations at different wind speeds, (**a**) nuclei mode, (**b**) Aitken mode, (**c**) accumulation mode, and (**d**) total particle – *Error bars* represent one standard deviation

$$E_N = \frac{\Delta[\mathrm{N}]}{\Delta[\mathrm{CO_2}] + \Delta[\mathrm{CO}]} W_c.10^{12} \tag{4.1}$$

where

E_N = emission factor for ultrafine particle (number count per kg of fuel burned);

$\Delta[N]$ = concentration of UFP above background level (number count per cm^3);

$\Delta[CO_2]$ and $\Delta[CO]$ = concentrations of CO_2 and CO (mg/m^3) above background level;

W_c = the weight fraction of carbon in the considered fuel (0.87 for diesel and 0.85 for gasoline) (Phuleria, Geller, Fine, & Sioutas, 2006).

This methodology can be used to characterize on-road motor vehicle emissions under real-world conditions. Because the vehicles observed in this study predominantly were gasoline-powered (99.5 % of the vehicles), 0.85 was adopted as the carbon weight fraction (Phuleria et al., 2006). The background concentrations for UFP, CO, and CO_2 at the sampling site were 1,359/cm^3, 0.0 ppm, and 399 ppm, respectively. The average particle emission factor throughout the sampling period of time was 2.36E+14 per kg fuel burned as shown in Fig. 4.8 (average vehicle speed was 20 km/h).

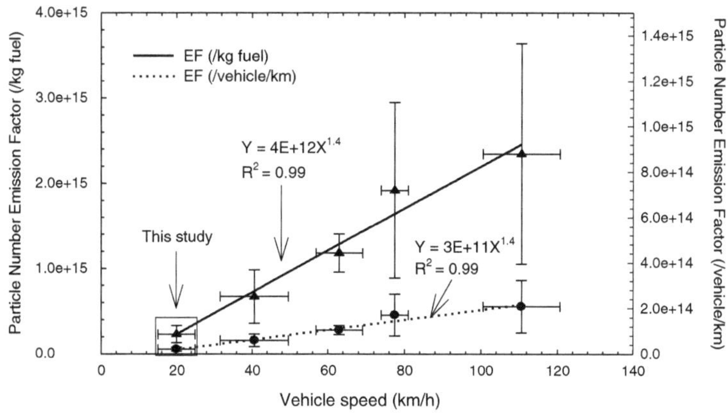

Fig. 4.8 Particle emission factors by vehicle speed – *Error bars* represent one standard deviation

It should be noted that at 19:00 the E_N was 1.43E+14 per kg fuel burned and it increased to 3.48E+14 per kg fuel burned at 20:00. Meteorological conditions between these 2 h were quite similar. The only difference observed in the field was the vehicle speed. During 19:00, a large amount of vehicles were waiting on the bridge to enter the U.S. Due to the security checking, all the vehicles had to wait at least half an hour on the bridge. While at 20:00, the traffic density on the bridge decreased to 25 % of the previous hour, which means vehicles could enter into the U.S. without idling on the bridge. Therefore, the relationship between vehicle speed and the particle emission factor needed to be determined. Some studies reported UFP emission rates in the unit of number count per km driven per vehicle. The following Eq. (4.2) was adopted to convert emission rate to fuel-based emission factors in the unit of number count per kg fuel burned (Ning, Polidori, Schauer, & Sioutas, 2008).

$$E_N = \frac{E_R}{C_f \times D_f} \tag{4.2}$$

Where

E_N = emission factor for UFP (number count/vehicle/km);
E_R = emission rate for UFP (number count per km driven per vehicle);
C_f = average fuel consumption rate (L/km);
D_f = fuel density (kg/L).

The average fuel consumption rate and fuel density for gasoline fuel used to calculate emission factors were 0.12 L/km and 0.74 kg/L (Owen & Coley, 1995). The previously determined emission factors for border crossing traffic were compared to other recently published data. Table 4.4 provides information on emission factors determined from chassis dynamometer and other on-road/near roadway studies. The emission factors for heavy-duty diesel vehicles are over one order of magnitude higher than the emission factors for gasoline-fueled vehicles (Morawska et al., 2005),

Table 4.4 Particle emission factors reported in the literature data for gasoline powered vehicles

Site description	Vehicle speed (km/h)	Particle size (nm)	$E_N \times 10^{14}$ (km^{-1})	$E_N \times 10^{14}$ (kg^{-1})	References
CA-110 Freeway	112	>10	1.60	18.0	Ning et al. (2008)
Caldecott tunnel, CA	59	>10	0.40	4.6	Kirchstetter, Harley, Kreisberg, Stolzenburg, and Hering (2002)
Chassis dynamometer	50	10–10,000	0.17	1.9	Farnlund, Holman, and Kageson (2001)
Highway, Minnesota	32	3–1,000	2.50	28.0	Kittelson et al. (2004)
Motorway, Germany	112	30–10,000	1.20	14.0	Corsmeier et al. (2005)
Suburban freeway, Australia	100	18–880	0.19	2.1	Morawska et al. (2005)
Busy street canyon, London	50	11–450	0.12	1.4	Jones and Harrison (2006)
Motorway	120	>7	6.90	77.0	Imhof et al. (2005)
	100		3.20	36.0	
	50		0.80	9.0	
Chassis dynamometer	40	7–400	0.11	1.2	Ristovski et al. (2004)
	60		0.10	1.1	
	80		0.09	1.0	
	120		0.20	2.2	
Chassis dynamometer	32	>3	0.19	2.2	Cadle et al. (2001)
Tunnel, Stockholm	70	3–900	2.70	30.0	Kristensson et al. (2004)
	75		3.30	37.0	

since the emission factors between diesel and gasoline fuels are not statistically comparable, the results from 11 prior studies presented in Table 4.4 only included assessment of emission factors for gasoline-fueled vehicles.

A summary of the results obtained from the comparison of average emission factors of diesel powered and gasoline powered vehicles as well as vehicle mix was presented by Morawska et al. (2005). The results showed that both dynamometer and on-road studies yielded comparable values for particle emission factors, and there was no statistical difference of the emission factors among vehicle fleets in different locations in the world. Therefore, the particle emission factors from different studies are equally suitable for quantification of the relationship between vehicle speed and emission factors.

The data presented in Table 4.4 are shown in Fig. 4.8. Those data were divided into four groups based on vehicle speed and two lines (solid and dotted) represent the particle emission factor in two different units (per kg fuel and per vehicle per km). Particle emission factors had significant correlations with vehicle speed

($R^2 = 0.99$). When the vehicle speed was below 30 km/h, the emission factors would be less than 4.7E+14 per kg fuel and 3.5E+13 per vehicle per km based on the two equations in Fig. 4.8 which were comparable to the measurement data in this study. Morawska et al. (2005) discovered that when the vehicular driving patterns, including acceleration, constant speed, deceleration, stop-go, or idling, varied over short distances, the emission factor for the gasoline-fueled vehicle was 2.4E+14 per kg fuel burned. The findings in this and previous studies (Table 4.4) demonstrate that the equations in Fig. 4.8 are applicable for gasoline-fueled vehicles at different traveling speeds.

4.4 Conclusions

In this study, UFP number size distribution showed strong temporal variations. The average number concentration of particles in 7–290 nm in the City of Laredo was 1.89×10^4 per cm^3. Direct vehicular emissions were the main source of UFP. Low UFP number concentrations were observed when the traffic densities were relatively low (during the evening of Christmas day and early morning after Christmas day). During nighttime, the particles in the size range 20–290 nm were mainly influenced by the effect of nucleation. BC concentration and UFP size distribution data collected during Christmas Day suggested that there could be some other underlying sources causing the increase in the UFP and BC mass concentrations, such as the biomass burning and fireworks in this region.

A regression analysis of the entire data set found a strong linear correlation between NO_x and the smaller particle size range 7–100 nm ($R^2 = 0.75$). Wind direction was found to have a significant effect on the UFP concentration and other pollutants. Correlations between particle number concentrations in the different size ranges and wind speed were calculated. Moderate to strong exponential decay trend was found between wind speed and number concentrations of particles in all four size ranges.

Significant correlation was found between UFP emission factors and vehicle speeds based on data collected in Laredo and those reported in the literature. Multivariate statistical analyses should be employed to better explain the effects of traffic and meteorological parameters (ambient temperature, relative humidity, wind direction, and wind speed) on UFP concentrations and size distributions. Receptor modeling analysis should be employed in further analysis to better characterize the sources of different pollutants.

Acknowledgements This material is based upon work supported by the Center for Research Excellence in Science and Technology – Research on Environmental Sustainability of Semi-Arid Coastal Areas (CREST-RESSACA) at Texas A&M University-Kingsville through funding from the National Science Foundation (NSF) under Cooperative Agreement No. HRD-0734850. The authors would also like to thank the City of Laredo and Texas Commission on Environmental Quality for their support and information provided for this research work. Any opinions, findings, and conclusions or recommendations expressed in this material are those of the authors and do not necessarily reflect the views of the National Science Foundation, the city of Laredo, or the Texas

Commission on Environmental Quality. The authors would like to express their sincere gratitude to Bob Castro, Joe Rogerio (City of Laredo), and other students from Texas A&M University–Kingsville for their help in conducting this research work.

References

Bai, N., Khazaei, M., Van Eeden, S., & Laher, I. (2007). The pharmacology of particulate matter air pollution-induced cardiovascular dysfunction. *Pharmacology and Therapeutics, 113*, 16–29.

Ban-Weiss, G., Lunden, M., Kirchstetter, T., & Harley, R. (2010). Size-resolved particle number and volume emission factors for on-road gasoline and diesel motor vehicles. *Journal of Aerosol Science, 41*, 5–12.

Bhugwant, C., Cachier, H., Bessafi, M., & Leveau, J. (2000). Impact of traffic on black carbon aerosol concentration at la Reunion Island (Southern Indian Ocean). *Atmospheric Environment, 34*, 3463–3473.

Cadle, S., Mulawa, P., Groblicki, P., Laroo, C., Ragazzi, R., Nelson, K., et al. (2001). In-use light-duty gasoline vehicle particulate matter emissions on three driving cycles. *Environmental Science and Technology, 35*, 26–32.

Chan, L., Chan, C., & Qin, Y. (1999). The effect of commuting microenvironment on commuter exposures to vehicles emission in Hong Kong. *Atmospheric Environment, 33*, 1777–1787.

Corsmeier, U., Imhof, D., Kohler, M., Kuhlwein, J., Kurtenbach, R., Petrea, M., et al. (2005). Comparison of measured and model-calculated real-world traffic emissions. *Atmospheric Environment, 39*, 5760–5775.

Diamond, R., & Parker, M. (2004). Preliminary air quality assessment related to traffic congestion at Windsor's ambassador bridge. In *Final report. Southwestern Region Ministry of the Environment* (p. 6). Ontario, Canada: Queen's Printer for Ontario.

Farnlund, J., Holman, C., & Kageson, P. (2001). Emissions of ultrafine particles from different types of light duty vehicles. In *Swedish National Road Administration Publication* (p. 16). London: ENDS Europe.

Gao, J., Wang, J., Cheng, S., Xue, L., Yan, H., Hou, L., et al. (2007). Number concentration and size distributions of submicron particles in Jinan urban area: Characteristics in summer and winter. *Journal of Environmental Sciences, 19*, 1466–1473.

Imhof, D., Weingartner, E., Ordonez, C., Gehrig, R., Hill, M., Buchmann, B., et al. (2005). Real-world emission factors of fine and ultrafine aerosol particles for different traffic situations in Switzerland. *Environmental Science and Technology, 39*, 8341–8350.

Jones, A., & Harrison, R. (2006). Estimation of the emission factors of particle number and mass fractions from traffic at a site where mean vehicle speeds vary over short distances. *Atmospheric Environment, 40*, 7125–7137.

Kerminen, V., Pakkanen, T., Makela, T., Hillamo, R., Sillanpaa, M., Ronkko, T., et al. (2007). Development of particle number size distribution near a major road in Helsinki during an episodic inversion situation. *Atmospheric Environment, 41*, 1759–1767.

Ketzel, M., Wahlin, P., Berkowicz, R., & Palmgren, F. (2003). Particle and trace gas emission factors under urban driving conditions in Copenhagen based on street and roof-level observations. *Atmospheric Environment, 37*, 2735–2749.

Kirchstetter, T., Harley, R., Kreisberg, N., Stolzenburg, M., & Hering, S. (2002). Corrigendum to on-road measurement of fine particle and nitrogen oxide emissions from light- and heavy-duty motor vehicles [(1999) *Atmospheric Environment, 33*(18), 2955–2968]. *Atmospheric Environment, 36*, 6059.

Kittelson, D. B. (1998). Engines and nanoparticles: A review. *Journal of Aerosol Science, 29*, 575–588.

Kittelson, D., Watts, W., & Johnson, J. (2004). Nanoparticle emissions on Minnesota highways. *Atmospheric Environment, 38*, 9–19.

Kristensson, A., Johansson, C., Westerholm, R., Swietlicki, E., Gidhagen, L., Wideqvist, U., et al. (2004). Real-world traffic emission factors of gases and particles measured in a road tunnel in Stockholm, Sweden. *Atmospheric Environment, 38*, 657–673.

Morawska, L., Jamriska, M., Thomas, S., Ferreira, L., Mengersen, K., Wraith, D., et al. (2005). Quantification of particle number emission factors for motor vehicles from on-road measurements. *Environmental Science and Technology, 39*, 9130–9139.

Morawska, L., Jayaratne, E., Mengersen, K., Jamriska, M., & Thomas, S. (2003). Differences in airborne particle and gaseous concentrations in urban air between weekdays and weekends. *Atmospheric Environment, 36*, 4375–4383.

Ning, Z., Polidori, A., Schauer, J., & Sioutas, C. (2008). Emission factors of PM species based on freeway measurements and comparison with tunnel and dynamometer studies. *Atmospheric Environment, 42*, 3099–3114.

Noble, C., Mukerjee, S., Gonzales, M., Rodes, C., Lawless, P., Natarajan, S., et al. (2003). Continuous measurement of fine and ultrafine particulate matter, criteria pollutants and meteorological conditions in urban El Paso, Texas. *Atmospheric Environment, 37*, 827–840.

Owen, K., & Coley, T. (1995). *Automotive fuels reference book* (2nd ed.). Warrendale, PA: Society of Automotive Engineers, Inc.

Palmgren, F., Wåhlin, P., Kildeso, J., Afshari, A., & Fogh, C. (2003). Characterization of particle emissions from the driving car fleet and the contribution to ambient and indoor particle concentrations. *Physics and Chemistry of the Earth, 28*, 327–334.

Phuleria, H., Geller, M., Fine, P., & Sioutas, C. (2006). Size-resolved emissions of organic tracers from light- and heavy-duty vehicles measured in a California roadway tunnel. *Environmental Science and Technology, 40*, 4109–4118.

Polichetti, G., Cocco, S., Spinali, A., Trimarco, V., & Nunziata, A. (2009). Effects of particulate matter (PM10, PM2.5 and PM1) on the cardiovascular system. *Toxicology, 261*, 1–8.

Pratt, G. C., Hendrickson, R. C., Chevone, B. I., Christopherson, D. A., O'Brien, M. V., & Krupa, S. V. (1983). Ozone and oxides of nitrogen in the rural upper-midwestern U.S.A. *Atmospheric Environment, 17*, 2013–2023.

Ristovski, Z., Morawska, L., Ayoko, G., Johnson, G., Gilbert, D., & Greenaway, C. (2004). Emissions from a vehicle fitted to operate on either petrol or compressed natural gas. *Science of the Total Environment, 323*, 179–194.

Sioutas, C., Delfino, R., & Singh, M. (2005). Exposure assessment for atmospheric ultrafine particles (UFPs) and implications in epidemiologic research. *Environmental Health Perspectives, 113*, 947–955.

Texas Department of Transportation. (2006). *Laredo AADT 2006*. Retrieved May 25, 2008, from http://www.dot.state.tx.us/apps/statewide_mapping/StatewidePlanningMap.html

United States Census Bureau. (2000). *U. S. census 2000*. Retrieved May 25, 2008, from http://factfinder.census.gov/servlet/GCTTable?_bm=y&-geo_id=04000US48_box_head_nbr=GCT-PH1&-ds_name=DEC_2000_SF1_U&-redoLog=false&-mt_name=DEC_2000_SF1_U_GCTPH1_ST2&-format=ST-7

Wang, Y., Hopke, P. K., Chalupa, D. C., & Utell, M. J. (2011). Long-term study of urban ultrafine particles and other pollutants. *Atmospheric Environment, 45*, 7672–7680.

Wang, Y., Zhu, Y., Salinas, R., Karnae, S., Ramirez, D., & John, K. (2008). Roadside measurements of ultrafine particles at a busy urban intersection. *Journal of the Air & Waste Management Association, 58*, 1449–1457.

Weijers, E., Khlystov, A., Kos, G., & Erisman, J. (2004). Variability of particulate matter concentrations along roads and motorways determined by a moving measurement unit. *Atmospheric Environment, 38*, 2993–3002.

Zhu, Y., Hinds, W., Kim, S., & Sioutas, C. (2002). Concentration and size distribution of ultrafine particles near a major highway. *Journal of the Air & Waste Management Association, 52*, 1032–1042.

Zunckel, M., Venjonoka, K., Pienaar, J., Brunke, E., Pretorius, O., Koosialee, A., et al. (2004). Surface ozone over southern Africa: Synthesis of monitoring results during the cross border air pollution impact assessment project. *Atmospheric Environment, 38*, 6139–6147.

Chapter 5
Evaluating the Impacts of Emissions and Global Temperature Perturbations on Air Quality in South and Central Texas

Neelesh Sule, Zuber Farooqui, Jhumoor Biswas, and Kuruvilla John

Abstract Recent regional-scale high ozone episodes over urban areas in South and Central Texas have prompted the Texas Commission on Environmental Quality (TCEQ) to designate the four urban regions of Austin, San Antonio, Corpus Christi, and Victoria as Near Non-attainment Areas (NNAs). These areas have to demonstrate an appropriate planning process for regulating the 8-h ozone concentrations by assessing the impact of emissions and designing effective control strategies. A base case modeling system was developed for the NNA regions using a high ozone episode during September 13–20, 1999. This system was then used to study the impact of various emissions categories and the influence of climate change induced atmospheric warming on ozone air quality in each of the NNAs. A greater ozone benefit was noted by reducing oxides of nitrogen (NO_x) emissions than volatile organic compounds (VOC) emissions in the NNAs. The simulations with increasing atmospheric temperatures showed significant impact on the 8-h ozone concentrations and its spatial extent in the NNA region. The analyses described

N. Sule
Providence Engineering & Environmental Group,
1200 Walnut Hill Lane, Irving, TX 75038, USA

Z. Farooqui
Department of Mechanical and Energy Engineering,
University of North Texas, Denton, TX 76203, USA

Current address: California Air Resources Board, Sacramento, CA 95814, USA

J. Biswas
Indian Institute of Social Welfare and Business Management (IISWBM),
Kolkata, West Bengal 700073, India

K. John (✉)
Department of Mechanical and Energy Engineering,
University of North Texas, Denton, TX 76203, USA
e-mail: kuruvilla.john@unt.edu

D. Ramirez et al. (eds.), *Environmental Sustainability Issues in the South Texas–Mexico Border Region*, DOI 10.1007/978-94-007-7122-2_5,

would help policy makers in developing robust emission control strategies needed to improve and maintain the air quality under changing climate and future emissions scenarios.

Implications. This study addresses the challenge of anthropogenic emission reductions under a changed climate scenario and its impact on urban and regional air quality management. Tools and analyses from this study can assist policy makers and air quality planners to develop strategies and guidance for effective environmental planning.

Keywords Photochemical modeling • Climate change • Urban air quality • Ozone concentration • Emissions impact

5.1 Introduction

Tropospheric ozone (O3) is one of the most widespread air pollution problems currently affecting major urban areas across the United States. Ozone is a secondary pollutant and it is formed non-linearly by a series of complex photochemical reactions involving its precursors, namely, oxides of nitrogen (NO_x) and volatile organic compounds (VOCs) under favorable meteorological conditions. The United States Environmental Protection Agency (U.S. EPA) has designated ozone as a criteria pollutant and has set stringent National Ambient Air Quality Standards (NAAQS). The regions which violate the federally mandated NAAQS for ozone are designated as nonattainment areas. Due to several recent regional-scale high ozone episodes that have impacted the urban areas in Texas, the Texas Commission on Environmental Quality (TCEQ) designated five urban regions as being Near Non-attainment Areas (NNAs) by a slim margin. An NNA is defined by TCEQ as a geographic area meeting NAAQS. These areas in Texas include the cities of Austin, Corpus Christi, Longview-Tyler-Marshall, San Antonio, and Victoria. According to the Clean Air Act (CAA), states with areas in non-attainment of the federal NAAQS for one or more criteria pollutants have to develop and submit a State Implementation Plan (SIP) to U.S. EPA. In Texas along with the non-attainment areas, the NNAs also have to demonstrate an appropriate planning process in attaining the 8-h NAAQS for ozone. Such a demonstration is often accomplished through the use of grid-based photochemical models, since they are designed to simulate the complex chemical and meteorological processes associated with the production and transport of ozone. Several earlier studies have illustrated the importance of these models in making regulatory decisions (Biswas, Hogrefe, Rao, Hao, & Sistla, 2001; Biswas & Rao, 2001; Hogrefe, Rao, Kasibhatla, Hao, et al., 2001; Hogrefe, Rao, Kasibhatla, Kallos, et al., 2001; Houyoux, Vucovich, Coats, Wheeler, & Kasibhatla, 2000; Kasibhatla & Chameides, 2000; Sistla et al., 2001). Since photochemical modeling for SIP development is costly and resource intensive, it is important to select a representative set of episodes and to not model all days that may exceed the 8-h standard (Durrenberger, Breitenbach, Red, & Sullivan, 1999).

Few studies have been conducted for the South Texas region (Biswas, John, & Farooqui, 2008, 2009; Farooqui, Biswas, & John, 2008) that evaluates the impact of emissions and climate change on the ambient ozone concentrations in the region. As per the climate model simulations conducted by scientists associated with the Intergovernmental Panel on Climate Change (IPCC), an average rise of global atmospheric temperature was predicted to be between 1.4 and 5.8 °C by the year 2100 (Meehl et al., 2007). It is widely known that high temperature and high ozone concentrations are strongly correlated. This is corroborated by other studies as well (Chen, Ho, Lai, & Chou, 2003; Sillman & Samson, 1995; Tecer, Erturk, & Cerit, 2003).

The development of a base case modeling system for the NNA regions in South and Central Texas using the September 13–20, 1999, high ozone episode is discussed in this study. The impact of reducing anthropogenic sources of emissions on the ambient 8-h ozone concentrations within each of the NNA regions is also highlighted in this work. Finally, to assess the impact of climate change on the ambient ozone concentrations in South and Central Texas, the changes in the chemistry of the atmosphere were evaluated for various perturbed temperature scenarios. The complexities associated with the confounding impacts of a warming atmosphere on the effectiveness of emissions control strategies on urban and regional air quality are explored in this study. Analyses such as these could potentially help policy makers in developing effective control strategies that would bring the region in attainment of the ozone NAAQS and also provide valuable information on the effects of climate variability on ambient ozone concentrations in the urban areas of Texas.

5.2 Modeling Domain

Three nested domains with spatial grid resolutions of 36, 12, and 4 km were used for photochemical modeling exercise for the high ozone episode of September 13–20, 1999, as shown in Fig. 5.1. The 4 km domain covered the four NNAs in South and Central Texas which include Austin, San Antonio, Victoria, and Corpus Christi. The NNA grid had 90 × 108 cells at 4 km resolution. The inset in Fig. 5.1 shows the four NNA regions along with the continuous air monitoring sites within each of the NNAs.

5.3 Model Inputs and Methodology

In order to maintain the urban air quality within the 8-h ozone NAAQS, each of the NNAs has to demonstrate an appropriate air quality planning process. This includes evaluation of the impacts of anthropogenic sources of emissions and the meteorological conditions on the generation and transport of ozone. Such a demonstration is often accomplished through the use of grid-based photochemical models, since they are designed to simulate the complex physical, chemical, and meteorological processes

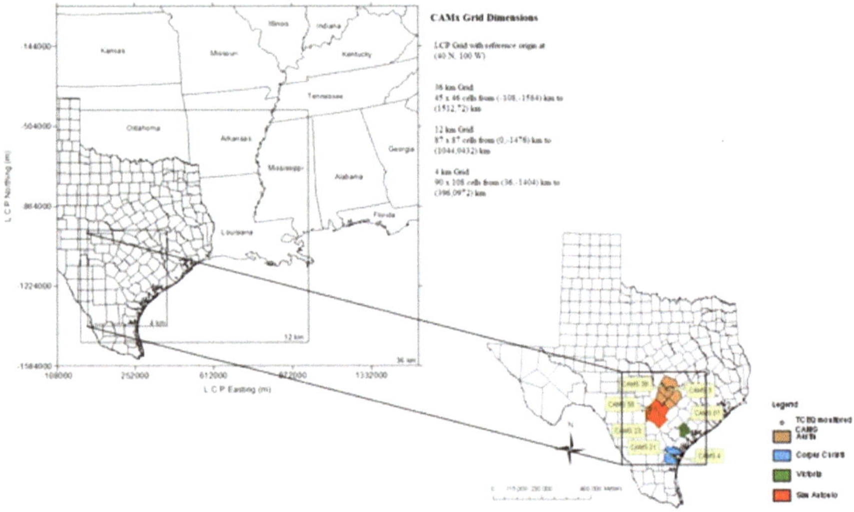

Fig. 5.1 CAMx modeling domains for the high ozone episode of September 1999

associated with the production of ozone. In this study, the meteorological fields were generated using the Fifth Generation Pennsylvania State University/National Center of Atmospheric Research (PSU/NCAR) Meteorological Model (MM5) (Grell, Dudhia, & Stauffer, 1995). The Emission Processing System version 2 (EPS2) (United States Environmental Protection Agency [U.S. EPA], 1992) was utilized to process emissions temporally and spatially to generate emissions input files. The photochemical model, Comprehensive Air quality Model with extension (CAMx) version 3.10 (ENVIRON International Corporation, 2002) was used in this study.

A single consistent modeling system was developed by TCEQ for all of the Texas NNAs. TCEQ and each of the NNAs analyzed the meteorological and measured ambient ozone data and collectively identified September 13–20, 1999, as the best representation of a single unifying ozone episode for all four NNA regions (Austin, San Antonio, Austin, and Corpus Christi). The first 3 days were considered to be ramp-up days and the episode days September 16–20, 1999, were selected to develop the base case for the NNA regions.

5.3.1 Model Inputs

Two key inputs required for photochemical models are emissions and meteorology. Photochemical models require specific processed and formatted emissions. Similarly, meteorology is also required to be processed as per photochemical model requirements. Processing and generation of required inputs are described in the following sections.

5.3.1.1 Meteorological Inputs

Meteorological fields for the air quality models are typically derived from either prognostic or diagnostic meteorological models. Meteorological models provide a realistic representation of the regional meteorological parameters with reduced uncertainty normally introduced by interpolation of errors as a result of sparse data from equally sparse observational networks. In the coastal region, the meteorology is usually characterized by local, thermally forced, daytime sea-breezes which advect emissions inland, bringing cold air from the sea-surface, forming an even shallower mixed layer (Cai & Steyn, 2000). Sea-surface temperature (SST) is a key component of the ocean-atmospheric system that drives convection over the open sea and induces sea-breeze circulation close to land, i.e., strength of inland penetration of sea-breeze fronts. It was found that the effect of spatially resolved SST data can influence meteorological model (MM5) predictions on coastal mesoscale circulation (Phadnis, Robe, Klausmann, & Scire, 2003). Substantial differences with the use of SST datasets are observed in predicting precipitation and wind patterns over the simulated domains. A study by Biswas and Rao (2001) revealed the importance of the ventilation co-efficient (product of wind speeds and mixing heights) in model simulated peak ozone concentrations on a regional basis. In addition, scaling down of modeled wind speeds in the Corpus Christi region was found to substantially improve the simulated peak ozone concentrations (John et al., 2005). Thus, it is important to simulate the coastal meteorological parameters as accurately as possible for South Texas. The meteorology for September 13–20, 1999, high ozone episode was developed by using MM5 version 3.4. The configuration of the MM5 for 1999 episode is summarized in Table 5.1.

5.3.1.2 Emissions Input

EPS2 was utilized to generate CAMx ready gridded emissions input files for the period of September 13–20, 1999. The emission inventories used in this study were from U.S. EPA's National Emissions Inventory 1999 (NEI99) version 1. Refined emissions developed by TCEQ were also used for the State of Texas. Emery, Tai, and Yarwood (2002) provide a comprehensive report on emissions inventory and emissions processing for the NNAs in Texas.

The relationship between ozone and its precursors is considered to be a major source of uncertainty which is further enhanced by the uncertainty in the emissions inventory. Prior studies have indicated that emissions are often underestimated (Sistla et al., 2001). Therefore, it is important to identify the emission sources and estimate emissions as accurately as possible. Thus, a comprehensive study of allocation and understanding of the distribution of emissions will be a first step towards the eventual development of successful emission reduction strategies. Figure 5.2 illustrates the weekday and weekend NO_x and VOC emissions in tons per day (tons/day) for all four NNAs. The variability in emissions on the weekdays

Table 5.1 Summary of MM5 configuration used for September 1999 high ozone episode

Parameters	Configuration used
Grid Sizes	Two-way interactive 108/36/12/4 km
Planetary boundary layer scheme	Medium Range Forecast (MRF)
Cumulus parameterization	Kain-Fritsch (except 4 km)
Radiation scheme	Rapid Radiative Transfer Model (RRTM)
Land soil model (LSM)	Five layer soil model
Four dimensional data assimilation (FDDA)	2-d surface analysis nudging
	3-d surface analysis nudging (only above boundary layer)
	Observation based nudging for 12 and 4 km

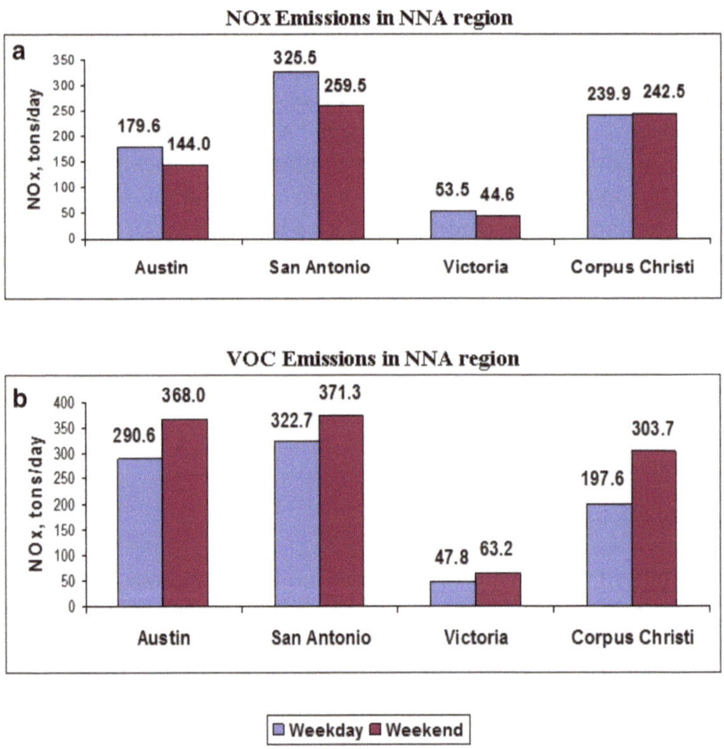

Fig. 5.2 Weekday/weekend emissions in NNA regions, (**a**) NO_x emissions tons/day and (**b**) VOC emissions tons/day

and weekend days were primarily due to variations in day-specific activities and to a lesser extent meteorological variations.

- **Austin**: In Austin, total weekday and weekend NO_x emissions were about 180 ton/day and about 144 ton/day, respectively. The NO_x emissions in Austin were mainly due to on-road mobile sources over the weekday and weekends, although the magnitude was lower over the weekends. The total VOC emissions were more than twice the NO_x emissions over the weekend with biogenic sources contributing the bulk of the emissions.

- **San Antonio**: The total NO_x emissions in San Antonio were highest amongst the four NNA regions with about 325 and 260 ton/day over the weekday and weekend, respectively. Similar to the Austin area the primary source of NO_x emissions were on-road mobile sources. San Antonio also had the maximum total VOC emissions of the four NNA regions with the primary source originating from biogenic emissions.
- **Victoria**: This region had the minimum total NO_x and VOC emissions among the four NNA regions. Unlike Austin and San Antonio, the primary sources of NO_x emissions in Victoria were elevated point sources in the region. However, similar to Austin and San Antonio the major contributor of VOC emissions were biogenic sources with total VOC emissions of about 48 and 63 ton/day for a weekday and weekend day, respectively.
- **Corpus Christi**: The emission distribution pattern in the Corpus Christi area was rather different when compared to the other urban areas within the study region. The weekday and weekend NO_x emissions were comparable with the non-road emission source categories dominating the NO_x emissions in the region. However, the VOC emissions in Corpus Christi were much higher on weekends than during weekdays. Since Corpus Christi is a port area and a sea resort, the emissions from non-road marine sources contribute significantly towards the total anthropogenic emissions (Farooqui, John, Biswas, & Sule, 2013). On weekends an increase in tourism-related activities resulted in higher non-road mobile emissions. Unlike other urban areas, biogenic emissions did not contribute significantly in this region. However, on weekends the total VOC emissions were larger than the total NO_x emissions due to increased anthropogenic VOC emissions from non-road and area source categories.

5.3.2 Methodology

5.3.2.1 Performance Evaluation of MM5 and CAMx for September 1999 Episode

Statistical evaluation of the MM5 model was performed for all available surface observation stations across San Antonio-Austin, Corpus Christi-Victoria, and across the 4 km domain. METSTAT program developed by ENVIRON, Inc. was used for the evaluation.

The program read MM5 output for wind speed, wind direction, temperature, and humidity to generate pairs of surface observation and model predictions. The following statistical metrics (ENVIRON International Corporation, 2004) were examined over the study region.

- Bias error: mean difference between pairings of predicted and observed data.
- Gross error: mean absolute value of difference between pairings of predicted and observed data.
- Root Mean Square Error (RMSE): the square root of the mean of the squared difference between pairings of predicted and observed data.
- Index of agreement (IOA): at each monitoring site, the sum of the absolute value of the difference between the prediction and the mean of the observations and the absolute value of the difference between the observation and the mean of the observations were calculated. These sums were added over all monitoring sites and divided into the square of the RMSE. This value was then subtracted from one.

Similarly, performance evaluation of the photochemical model CAMx was conducted as per EPA recommended statistical methods since the acceptance of the developed photochemical models was based on its performance. The following statistical metrics were analyzed to evaluate the base case model performance results:

- Unpaired Peak Accuracy (UPA): the relative difference in percentile between the observed peak and the estimated peak among all sites within a given distance from the observed peak site at any time. EPA suggests that the result from this test should be within ± 20 %.
- Average Paired Peak Accuracy (APPA): the average value of the relative difference in percentile between the observed peak that was greater than 60 parts per billion (ppb) and the estimated peak at each site.
- Peak Timing Bias (PTB): the average value of the difference in hours between the time of the observed peak that was greater than 60 ppb and the time of the estimated peak at each site.
- Overall Bias (OB): the average value of the relative difference in percentile between all observed values that were greater than 60 ppb and all estimated values at each site and hour. The EPA suggests that the result from this test should be within ± 15 %.
- Overall Gross Error (OGE): the average value of the absolute relative difference in percentile between all observed values that were greater than 60 ppb and all estimated values at each site and hour. EPA suggests that the result from this test should be below 35 %.

5.3.2.2 Analysis of the Impact of Major Emission Source Categories

The potential control strategies for a region can be resolved only after the influence from various emission source categories has been ascertained. This was accomplished by zeroing out the emissions from different source categories and determining the reduction in ozone concentrations as a consequence. A number of additional emission sensitivity runs were performed by zeroing out the emissions from elevated point, mobile, biogenic, and non-road emissions categories for the different NNA regions to accurately assess the impact of different sources of emissions and subsequently help in the development of potential emission control strategies for each of the NNA region. The influence of a particular emissions-reduction run in a region was estimated by analyzing the reduction in the ozone peaks during the ozone episode. The positive differences revealed ozone benefit and negative differences showed ozone dis-benefits.

5.3.2.3 Impact of Atmospheric Warming on Urban and Regional Ozone

The impact of temperature changes was evaluated by conducting sensitivity simulations using the September 1999 base case that was developed earlier. The surface and upper air temperatures were varied from 2 to 6 °C to replicate the anticipated changes in global temperature according to the IPCC report (Meehl et al., 2007). This study investigated the spatial impact of atmospheric temperature changes on the peak ozone concentrations and spatial distribution patterns of surface level ozone concentrations.

5.4 Results and Discussion

5.4.1 Performance Evaluation for September 1999 Episode

5.4.1.1 Performance Evaluation of MM5

Statistical performance evaluation of the MM5 model was performed for the combined Corpus Christi and Victoria areas, San Antonio and Austin areas, and for the entire 4 km domain covering all the NNA regions. The results of the model performance evaluation are summarized in Table 5.2. It was found that the MM5 model performed very well for wind speed and humidity. For these parameters the model also performed within the limits for the Corpus Christi and Victoria areas but did not do as well in the Austin and San Antonio areas. However, the wind directions modeled for the Corpus Christi and Victoria area were outside the limits. The overall performance of the model was acceptable for its use in CAMx photochemical modeling as recommended by EPA (U.S. EPA, 2005).

5.4.1.2 Performance Evaluation of CAMx

The CAMx was set up to simulate an 8 day period in September 1999. However, the base case evaluation was done only for the high ozone days of September 16–19, 1999, since the initial days were used as model ramp-up days. The results of the model performance evaluation are summarized in Table 5.3. The UPA values were

Table 5.2 Statistical analyses of MM5 run for the September 13–20, 1999, high ozone episode

	Limit	Corpus Christi/Victoria	San Antonio/Austin	NNA Domain
Wind speed bias	±0.5 m/s	−0.1	−0.1	0.3
Wind speed RMSE	<2 m/s	1.2	1.1	1.5
Wind speed IOA	>0.6	0.7	0.7	0.8
Wind direction bias	±10 deg	−2.4	2.7	−1.8
Wind direction error	<30 deg	30.1[a]	25.1	31.9[a]
Temperature bias	±0.5 K	0.5	−1.3[a]	−0.5
Temperature error	<2 K	1.3	1.7	1.5
Temperature IOA	>0.8	1.0	1.0	1.0
Humidity bias	±1 g/kg	0.0	0.7	0.4
Humidity error	<2 g/kg	1.7	1.6	1.4
Humidity IOA	>0.6	0.7	0.5	0.9

[a]Indicates value above the permissible limit for performance evaluation

Table 5.3 Model performance evaluation statistics for the September 16–19, 1999, high ozone episode

	16-Sept	17-Sept	18-Sept	19-Sept
San Antonio				
UPA, %	−10	−5.4	−6.2	4.3
APPA, %	−22.9	−16.8	−19.1	−11.5
PTB, h	1	2	0	0
OB, %	−19.2[a]	−13.9	−18.8[a]	−9.2
OGE, %	19.2	13.9	18.8	11.5
Austin				
UPA, %	−16.9	−17.6	−21	−18.1
APPA, %	−20.1	−19.8	−22.3	−18.5
PTB, h	1	0	1	1
OB, %	−19.2[a]	−21.4[a]	−23.4[a]	−16[a]
OGE, %	19.2	21.4	23.4	16
Victoria				
UPA, %	−0.2	−1.6	−10	−10.5
APPA, %	−13.8	−15.7	−25.8	−16.2
PTB, h	0	1	−1	0
OB, %	−12.6	−17.5[a]	−28.5[a]	−26.5[a]
OGE, %	12.6	17.5	28.5	26.5
Corpus Christi				
UPA, %	−16.1	−15	−9.5	−6.8
APPA, %	−32.6	−31.2	−19.4	−19.4
PTB, h	0	−2	−1	2
OB, %	−33.3[a]	−29.4[a]	−22.6[a]	−19.4[a]
OGE, %	33.3	29.4	22.6	19.4

[a]Indicates value above the permissible limit for performance evaluation

Table 5.4 Number and spatial extent of 8-h ozone exceedances

Ozone concentrations	8-h ozone exceedance >74 ppb	Grid cell exceedance >74 ppb
Austin	1,541	368
San Antonio	1,156	206
Corpus Christi	416	122
Victoria	264	102

within the ±20 % limits set by EPA with mixed (both positive and negative) values observed for San Antonio, Austin, Victoria, and Corpus Christi. The PTB was 2 h on September 17 for San Antonio, while Austin showed a maximum bias error of 1 h. Victoria and Corpus Christi both showed a mixed timing bias. Overall, the base case model developed under-predicted the observed peak ozone levels throughout the study period in most areas. The overall pattern of diurnal variability in ozone concentrations was quite similar between the observation and the base case. Despite the under prediction, the model did reasonably well for each of the urban areas and was within the U.S. EPA prescribed acceptable norms of model performance.

As an integral part of the base case evaluation of the episode, ozone exceedance analysis was also conducted where the total number of the 8-h averaged peak ozone and grid exceedances over the NNA regions were calculated based on the 8-h ozone NAAQS (75 ppb). Table 5.4 summarizes the 8-h averaged ozone and grid exceedances. With the recent strengthening of the ozone standard to 75 ppb, all NNA regions showed an increase in the total number of 8-h averaged peak ozone and the number of grid cell exceedances.

5.4.2 *Analysis of the Impact of Major Emission Source Categories*

The analysis of the impact of reducing various emission source categories will typically provide guidance on designing effective control strategies for any region. Figure 5.3 provides the graphical representation of the results for each area. The results of this analysis are discussed for each of the NNAs.

- **Austin**: The on-road mobile source emissions in Austin had the greatest impact on the ground level ozone, by contributing up to 9.8 ppb of the peak 8-h ozone levels followed by point sources with peak ozone reductions of up to 8.3 ppb. The zero-out runs for area and non-road sources had relatively lower impact of 3.5 ppb each on the 8-h peak ozone concentrations. In Austin, about 100 ton/day of the NO_x emissions were from on-road mobile sources; thus, this source category had the largest impact on the 8-h ozone concentrations. Overall, the anthropogenic emissions in Austin were dictated by NO_x emissions. Therefore reducing NO_x emissions would result in a greater decrease in the peak ozone levels.

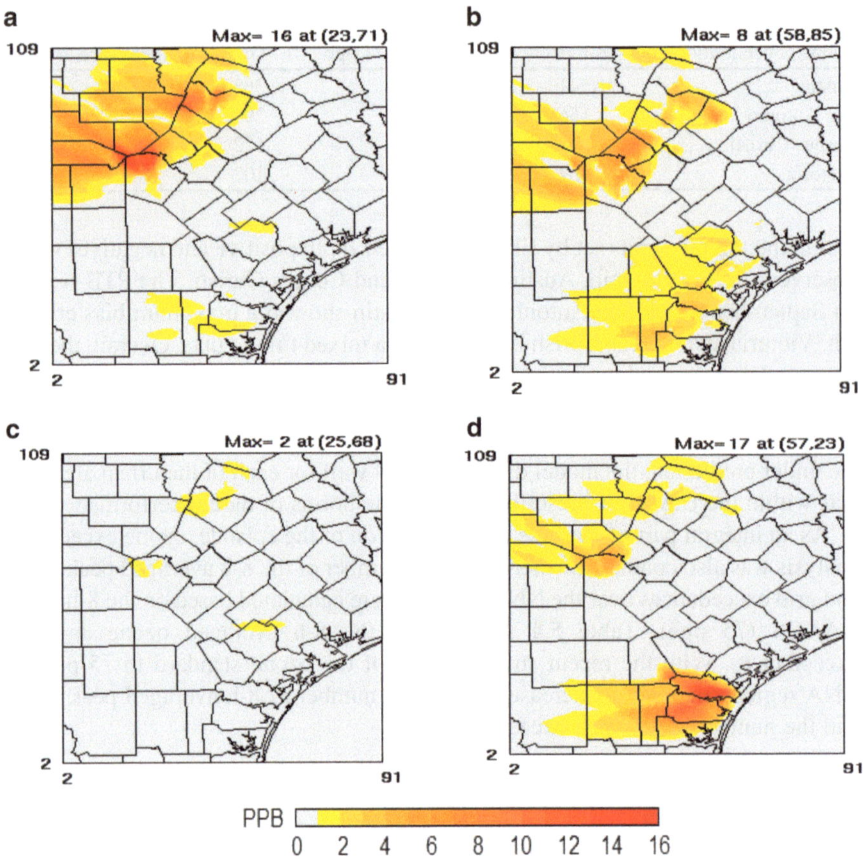

Fig. 5.3 Impact analysis of emissions sources in NNA region for (**a**) on-road mobile source, (**b**) point source, (**c**) area source, and (**d**) non-road source

- **San Antonio**: **Similarly, San Antonio was also** strongly influenced by on-road mobile source emissions which contributed about 150 ton/day of NO_x emissions on a weekday. Zero-out runs showed a reduction of 15.5 ppb in the ozone maxima. Further, zeroing-out point and non-road emission sources resulted in the reduction of peak ozone concentrations by 8 and 6.5 ppb, respectively. Overall, the elimination of total anthropogenic emissions resulted in a net decrease of 30 ppb in the peak ozone value for this area.
- **Corpus Christi**: **In Corpus Christi, unlike Austin and San Antonio**, the non-road source category contributed the largest amounts of NO_x and VOC emissions. The zero-out of non-road mobile emissions caused a decrease of 17 ppb in the peak values of ozone. The zero-out runs for other emission sources such as area, on-road mobile, and point sources had an impact of less than 2 ppb in reducing the maximum 8-h ozone concentrations. Overall, the anthropogenic

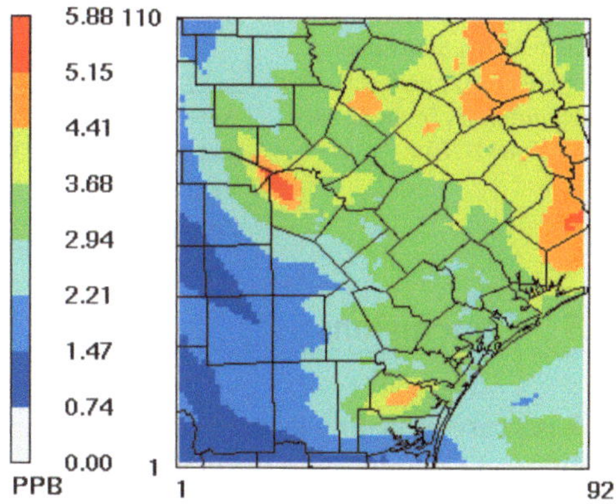

Fig. 5.4 Spatial distribution of 8-h ozone concentrations between the base case and the perturbed temperature (4 °C) case

emissions dominated by non-road emission sources were responsible for around 20 ppb of the predicted peak ozone levels.

- **Victoria**: This urban area was predominantly influenced by elevated point source emissions which contributed about 20 ton/day of NO_x emissions. The zero-out runs of the point sources showed a 3 ppb reduction in the ozone maxima while the zero-out runs for area, on-road mobile, and non-road sources impacted less than 2 ppb each in the predicted 8-h ozone. Overall, the anthropogenic emissions within this airshed contributed 4 ppb to the net 8-h averaged peak ozone concentrations.

5.4.3 Impact of Atmospheric Warming on Urban and Regional Ozone

As per the IPCC report, an average rise of global temperature was predicted to be between 1.4 and 5.8 °C by the year 2100 (Meehl et al., 2007). Therefore, an average rise in temperature of 4 °C was considered for this study. The spatial impact of temperature change of 4 °C on 8-h ozone concentrations over the NNA region is shown in Fig. 5.4. About a 6 ppb increase in ozone was noted in San Antonio and in the counties downwind of the Houston-Galveston area for a corresponding 4 °C increase in the ambient temperature. In Austin the maximum ozone concentration would increase by about 5 ppb while in Corpus Christi the maximum change in ozone concentration would be about 4.5 ppb.

The ozone concentrations predicted in each of the four NNAs for increases in temperature from 2 to 6 °C are summarized in Table 5.5. The maximum increase in the episodic 8-h ozone concentrations was observed to be at 6 °C in the San Antonio region. The regions downwind of Houston-Galveston had an increase of 8.53 ppb followed by increases in Austin, Corpus Christi, and Victoria region as reported by Biswas, John, and Farooqui (2009).

Table 5.5 Maximum impact of increase in atmospheric temperature on ozone concentrations (ppb) at each of the four NNAs

	Maximum impact of increase in temperature				
	+2 °C	+3 °C	+4 °C	+5 °C	+6 °C
Austin	2.65	3.85	5.12	6.24	7.38
San Antonio	3.04	4.49	5.88	7.24	8.53
Victoria	1.92	2.70	3.50	4.43	4.97
Corpus Christi	2.42	3.60	4.69	5.74	6.74

5.5 Conclusions

A base-case modeling system using CAMx was developed for the NNA regions in South and Central Texas using a September 13–20, 1999, high ozone episode. The model used MM5 derived meteorology along with EPA's NEI99 and TCEQ refined emissions for Texas. The overall model performance was within the EPA prescribed acceptable norms for each of the NNAs.

An assortment of ozone sensitivity runs were accomplished by reducing major emission categories over the urban regions in South and Central Texas for the high ozone days of September 16–19, 1999. The results conclusively indicated that emission distribution patterns strongly influenced the directionality of emission controls. There was considerable variability in the dominance of different emission source categories with their consequent impacts on peak ozone levels in different urban areas. The results from zero-out emission runs indicated an initial preliminary guide for the type of emission reductions needed in the urban areas of South and Central Texas. A majority of these urban areas revealed a greater ozone benefit from NO_x reductions than from VOC reductions.

The ozone concentrations in San Antonio and Austin were influenced mostly by emissions from on-road sources, and Victoria was significantly impacted by elevated point source emissions. The results showed that Corpus Christi was influenced by non-road emissions primarily ocean going vessels (OGV), barges, and pleasure-crafts with a net impact of 3.4 ppb on the peak 8-h ozone over the airshed.

Finally, to evaluate the impact of climate change on the air quality, photochemical model sensitivity runs accounting for temperature perturbations were conducted over the NNA region in South Texas using the September 1999 high ozone episode. The results clearly indicate that the rise in global temperature would have a significant impact on the 8-h ozone concentrations and the spatial extent of the ozone exceedances in the NNA region. Atmospheric warming as a result of climate change is extremely important for areas that are currently in non-attainment and near non-attainment of the 8-h ozone standard, since any further increase in atmospheric temperature would increase the amount and extent of ozone exceedances. This study demonstrates the need for accounting for the influence of the incorporation of climate change while designing effective emission control strategies for air quality management.

Acknowledgments This material is based upon work supported by the Center for Research Excellence in Science and Technology – Research on Environmental Sustainability of Semi-Arid Coastal Areas (CREST-RESSACA) at Texas A&M University-Kingsville through Cooperative Agreements (No. HRD-0206259 and HRD-0734850) from the National Science Foundation (NSF). Any opinions, findings, and conclusions or recommendations expressed in this material are those of the authors and do not necessarily reflect the views of the National Science Foundation.

References

Biswas, J., Hogrefe, C., Rao, S., Hao, W., & Sistla, G. (2001). Evaluating the performance of regional-scale photochemical modeling systems. Part III: Precursor predictions. *Atmospheric Environment, 35*, 6129–6149.

Biswas, J., John, K., & Farooqui, Z. (2008). Modeling the impact of temperature perturbations on urban and regional ozone levels in Texas. In *Proceedings of the 19th IASTED international conference on Modelling and Simulation*. Quebec City, QC: ACTA Press.

Biswas, J., John, K., & Farooqui, Z. (2009). Impact assessment of global temperature perturbations on urban and regional ozone levels in South Texas. *Advances in Geosciences: Atmospheric Science, 10*, 197–211.

Biswas, J., & Rao, S.T. (2001). Uncertainties in episodic ozone modeling stemming from uncertainties in the meteorological fields. *Journal of Applied Meteorology, 40*, 117–135.

Cai, X., & Steyn, D. (2000). Modeling study of sea breezes in a complex coastal environment. *Atmospheric Environment, 34*, 2873–2885.

Chen, K., Ho, Y., Lai, C., & Chou, Y. (2003). Photochemical modeling and analysis of meteorological parameters during ozone episodes in Kaohsiung, Taiwan. *Atmospheric Environment, 37*, 1811–1823.

Durrenberger, C., Breitenbach, P., Red, J., & Sullivan, D. (1999). Development of a conceptual model for episode selection of high eight-hour ozone events in the Dallas/Fort Worth area. In *Proceeding of 92nd Annual Air & Waste Management Association's Conference*, St. Louis, MO.

Emery, C., Tai, E., & Yarwood, G. (2002). *Development of a joint CAMx photochemical modeling episode for the four southern Texas near non-attainment area*. Novato, CA: ENVIRON International Corporation.

ENVIRON International Corporation. (2002). *User's guide comprehensive air quality model with extensions (CAMx) version 3.10*. Novato, CA: ENVIRON.

ENVIRON International Corporation. (2004). *Development of the September 13–20, 1999 base case photochemical model for Austin's early action compact*. Novato, CA: ENVIRON.

Farooqui, Z., Biswas, J., & John, K. (2008). Potential impact of climate perturbations on ozone episodes over an urban area in South Texas. In *Proceedings of the 101st Annual Air & Waste Management Conference*, Portland, OR.

Farooqui, Z. M., John, K., Biswas, J., & Sule, N. (2013). Modeling analysis of the impact of anthropogenic emission sources on ozone concentration over selected urban areas in Texas. *Atmospheric Pollution Research, 4*, 33–42.

Grell, G., Dudhia, J., & Stauffer, D. R. (1995). *A description of the fifth-generation Penn-State/NCAR mesoscale model (MM5)* (NCAR Tech Note NCAR/TN-398 +STR, 117 pp).

Hogrefe, C., Rao, S., Kasibhatla, P., Hao, W., Sistla, G., Mathur, R., et al. (2001). Evaluating the performance of regional-scale photochemical modeling systems. Part II: Ozone predictions. *Atmospheric Environment, 35*, 4175–4188.

Hogrefe, C., Rao, S., Kasibhatla, P., Kallos, G., Tremback, C., Hao, W., et al. (2001). Evaluating the performance of regional-scale photochemical modeling systems, part I: Meteorological predictions. *Atmospheric Environment, 35*, 4159–4174.

Houyoux, M., Vucovich, J., Coats, C., Jr., Wheeler, N., & Kasibhatla, P. (2000). Emission inventory development and processing for the seasonal model for regional air quality (SMRAQ) project. *Journal of Geophysical Research, 105*, 9079–9090.

John, K., Jung, I., Farooqui, M. Z., Kim, M., Marek, D., & Varghese, R. (2005). *Rider13* – Final report (Submitted to Texas Commission on Environmental Quality, Austin, TX). Kingsville, TX: Texas A&M University-Kingsville.

Kasibhatla, P., & Chameides, W. (2000). Seasonal modeling of regional ozone pollution in the eastern United States. *Geophysical Research, 27*, 1415–1418.

Meehl, G., Stocker, T., Collins, W., Friedlingstein, P., Gaye, A., Gregory, J., et al. (2007). Global climate projections. In S. Solomon, D. Qin, M. Manning, Z. Chen, M. Marquis, K. Averyt, M. Tignor, & H. Miller (Eds.), *Climate change 2007: The physical science basis – Contribution of working group I to the fourth assessment report of the intergovernmental panel on climate change*. New York: Cambridge University Press.

Phadnis, M., Robe, F., Klausmann, M., & Scire, J. (2003). Importance of the spatial resolution of sea-surface temperature data in meteorological modeling. In *13th PSU/NCAR Mesoscale Model user's workshop*. Boulder, CO: National Center for Atmospheric Research.

Sillman, S., & Samson, P. J. (1995). Impact of temperature on oxidant photochemistry in urban, polluted rural and remote environments. *Journal of Geophysical Research, 100*, 14175–14188.

Sistla, G., Hao, W., Ku, J., Kallos, G., Zhang, K., Mao, H., et al. (2001). An operational evaluation of the two regional-scale ozone air quality modeling systems over the eastern United States. *Bulletin of the American Meteorological Society, 82*, 945–964.

Tecer, L. H., Erturk, F., & Cerit, O. (2003). Development of a regression model to forecast ozone concentrations in Istanbul City, Turkey. *Fresenius Environmental Bulletin, 12*, 1133–1143.

United States Environmental Protection Agency [U.S. EPA]. (1992). *User's guide for the urban airshed model volume IV: User's manual for the emissions processor system 2.0* (EPA-450/4-90-007D(R)). Research Triangle Park, NC: U.S. Office of Air Quality and Planning Standards.

U.S. EPA. (2005). *Guidance on the use of models and other analysis in attainment demonstrations for the 8-hour ozone NAAQS* (EPA-454/R-05-002). Research Triangle Park, NC: U.S. Office of Air Quality Planning and Standards Emissions, Monitoring, and Analysis Division.

Chapter 6
Manufacture and Physical Characterization of Wood-Derived Activated Carbon from South Texas Mesquite for Environmental Applications

Amrutha Sakaray and David Ramirez

Abstract Activated carbons are widely used to remove contaminants from waste streams. Desirable properties of activated carbon include large surface area, porosity, high degree of selectivity between contaminant and carrier medium, rapid transport of the contaminant from bulk stream to the internal porous structure of the adsorbent, high adsorption capacity at low contaminant concentration, and available commercially at competitive prices. This research focuses on the production and physical characterization of carbonaceous adsorbents from low cost mesquite woodchips available in the South Texas region. Activated carbons were prepared from mesquite woodchips in a bench-scale tubular reactor. Steam and a potassium carbonate solution were used for the physical and chemical activation methods to manufacture activated carbons. The activated carbons were then characterized in terms of their physical and adsorption capacities for nitrogen at 77 K and methanol and water vapor at ambient temperature. The surface areas of the samples were determined using the N_2-Brunauer-Emmett-Teller method. The percent yield, bulk density, pore size distribution, average pore width, and total pore volume were evaluated. Equilibrium adsorption capacities of methanol and water vapor were determined by a gravimetric method. Experimental adsorption capacity data were in agreement with the Freundlich and Langmuir adsorption models.

A. Sakaray
Department of Environmental Engineering, Frank H. Dotterweich College of Engineering, Texas A&M University-Kingsville, Kingsville, TX 78363-8202, USA

FutureNet Group, Inc, Detroit, MI, USA

D. Ramirez (✉)
Department of Environmental Engineering, Frank H. Dotterweich College of Engineering, Texas A&M University-Kingsville, Kingsville, TX 78363-8202, USA

Center for Research Excellence in Science and Technology – Research on Environmental Sustainability of Semi-Arid Coastal Areas (CREST-RESSACA), Texas A&M University-Kingsville, Kingsville, TX 78363-8202, USA
e-mail: kfdr000@tamuk.edu

D. Ramirez et al. (eds.), *Environmental Sustainability Issues in the South Texas–Mexico Border Region*, DOI 10.1007/978-94-007-7122-2_6,

Keywords Adsorption • Carbonization • Activation • Air pollution • Mesquite tree

6.1 Introduction

Activated carbon is extensively used in adsorption processes to remove and recover or destroy hazardous pollutants from gas and water streams (Ramirez, Sullivan, Rood, & Hay, 2004; Suhas, Carrott, & Carrott, 2007; Sullivan et al., 2004). Activated carbons are highly porous with extended surface area and are available with a wider range of morphologies that provide desirable pollutant adsorption and regeneration properties (Luo et al., 2006; Suhas et al., 2007; Sullivan, Stone, Hashisho, & Rood, 2007). Adsorbents possess a porous structure and have a tangled network which enables them to capture pollutants that take on free unoccupied pores which are available within the activated carbon (Avelar, Bianchi, Goncalves, & Gaspar da Mota, 2010). Characterization of activated carbons is important because it better describes the physical, electrical, and adsorption properties of the adsorbents with specific morphologies which are used to capture or recover organic vapors from gas streams (Bansal, Donnet, Stoeckli, & Dekker, 1988). Activated carbon materials are also used for catalyst support, double-layer capacitors, battery electrodes, gas storage, adsorbents for bulky molecules, and in the bio-medical engineering field (Lin & Teng, 2003; Rosas, Bedia, Mirasol, & Cordero, 2008).

Activated carbons can be manufactured from carbonaceous precursors including wood, coal, coconut shells, lignin, waste rubber, poultry litter, leather wastes, and some polymers (Adinata, Daud, & Aroua, 2005; Avelar et al., 2010; Baccara, Bouzida, Fekib, & Montiela, 2009; Hayashi, Kazehayaa, Muroyama, & Watkinson, 2000; Kantarli & Yanik, 2010; Khezami, Chetouani, Taouk, & Capart, 2005; Lehmann, Rostam-Abadi, Rood, & Sun, 1998; Lima & Marshall, 2005; Qui & Guo, 2009; Suhas et al., 2007; Wua, Tsengb, Hu, & Wang, 2004). The production of activated carbon adsorbents from South Texas native mesquite woodchips can provide a two-fold environmental and economic benefit: A recycling path is created for waste wood materials, and novel low-cost engineered adsorbents are produced with physical and chemical activation methods for use in air quality applications. Mesquite plants are tree or thorny shrubs of the legume family growing naturally in arid and semi-arid areas of North and South America, Eastern Asia, and Northern Africa. The two species of interest in the Southwestern United States are the honey mesquite and the velvet mesquite. Honey mesquite is abundantly found in the South Texas region, New Mexico, and Northern Mexico.

Activated carbons can be granular, powdered, spherical, monolith, or in fabric cloth (Bansal et al., 1988). Granular activated carbon (GAC) can be made from coconut shells, bituminous coal, petroleum coke, or lignite. GAC can have an internal surface area ranging from 500 to 2,000 m^2/g (Bansode, Losso, Marshall, Rao, & Portier, 2004). Powdered activated carbon (PAC) is used primarily for removal of undesired smell, color, and other impurities in water streams. Highly mesoporous spherical activated carbon (SAC) can be prepared from divinylbenzene-derived polymers. SACs offer strengthened mechanical properties, high adsorption efficiency, improved resistance to abrasion, and low resistance to liquid diffusion (Wua et al., 2004). Activated carbon monolith is prepared from PAC. It is a coal-based adsorbent with a specific surface area ranging from 900 to 1,100 m^2/g and offers minimum pressure drop (Luo et al., 2006). Activated carbon fiber cloth, made from natural or synthetic fibers, is predominantly micropore (>90 %) with mean total pore width ranging from 7.4 to 9.9 Å, high specific surface area, and fast adsorbing and desorbing rates (Sullivan et al., 2007; Yanhe, Xie, Shuo, Shibo, & Yaobin, 2007). The properties most desirable in an activated carbon include high adsorption capacity for gaseous contaminants, ease of regeneration, totally desorbed, high mass transfer of pollutants to and from the activated carbon in gas or liquid phase adsorption keeping its adsorption capacities unchanged, and being chemically stable (Yanhe et al., 2007).

Production of activated carbon involves the carbonization and the activation processes of the carbonaceous material at high temperatures using an inert carrier gas. The *carbonization process* is the thermal decomposition of the carbonaceous material eliminating non-carbon species and producing a fixed carbon mass and a rudimentary pore structure. The operation is usually performed in a continuous stream of nitrogen gas in a furnace at temperatures below 800 °C (Sudaryanto, Hartono, Irawaty, Hindarso, & Ismadji, 2006). Carbonization and activation of coal can result in a primarily micro-porous adsorbent with specific surface areas approaching 1,000 m^2/g and moderate pore volumes (Bansal et al., 1988; Lehmann et al., 1998). The most important parameters which determine the quality and the yield of carbonized product are the final temperature, rate of heating, soaking time at the final temperature, and the nature and physical state of raw material (Hayashi et al., 2000). The material porosity is enhanced during the a*ctivation process*. There are two types of activation: chemical and physical. The most commonly used chemical activating agents include zinc chloride, phosphoric acid, potassium thio-cyanate hydroxides, sulfuric acid, and potassium carbonate (Baccara et al., 2009; Bansal et al., 1988). The process in which oxidizing gases, like carbon dioxide or steam, are used for the activation is called physical activation. The main objective of the activation process is to enhance the volume, enlarge the diameters of the pores, and create new porosity. It has been suggested that the physical method is the preferred activation method because it is more environmentally friendly and less costly as compared to chemical methods (Lehmann et al.).

In this study, mesquite-derived activated carbon was prepared by two processes: carbonization followed by activation. The carbonization process was carried out at temperatures ranging from 600 to 800 °C. The activation processes occur when using steam and chemicals such as a potassium carbonate solution. Generally, chemicals which are used in the preparation of activated carbon materials are zinc chloride, phosphoric acid, potassium hydroxide, and sodium hydroxide (Lua & Yang, 2004). However, alkali hydroxides such as potassium hydroxide, zinc chloride, and sodium hydroxide are hazardous, expensive, and corrosive chemicals that can be non-environmentally friendly and create waste disposal problems.

The manufactured activated carbons were characterized in terms of their physical and sorption properties. Physical properties included surface area, percent yield, bulk density, average pore width, pore size distribution, and total pore volume.

The effects of the carbonization temperature and activation time on the physical properties of activated carbon were assessed. The Freundlich and Langmuir adsorption models were used to model the experimental data.

6.2 Methods

6.2.1 *Experiment*

The experimental set up consisted of compressed ultra-high purity nitrogen (UHP-N_2) as the carrier gas, a steam generation system, a three-zone temperature controller furnace reactor, and a temperature monitor with a data logger system for data acquisition (Fig. 6.1). Mesquite samples were carbonized in a horizontal tubular furnace. Chemical activation of the samples was carried out with a solution of potassium carbonate. The temperature was measured using a thermocouple (Omega, K-type) placed inside the reactor above the sample crucible. In the physical activation process, the samples were activated by steam. The steam was generated using a syringe pump which injected deionized water, a stainless steel tube wrapped with heating tape to provide the desired steam temperature, and a heat controller.

The manufacture of the activated carbon samples involved two main processes: carbonization and chemical or physical activation of the carbonaceous material at high temperatures using UHP-N_2 as the inert carrier gas. Preparation of the sample,

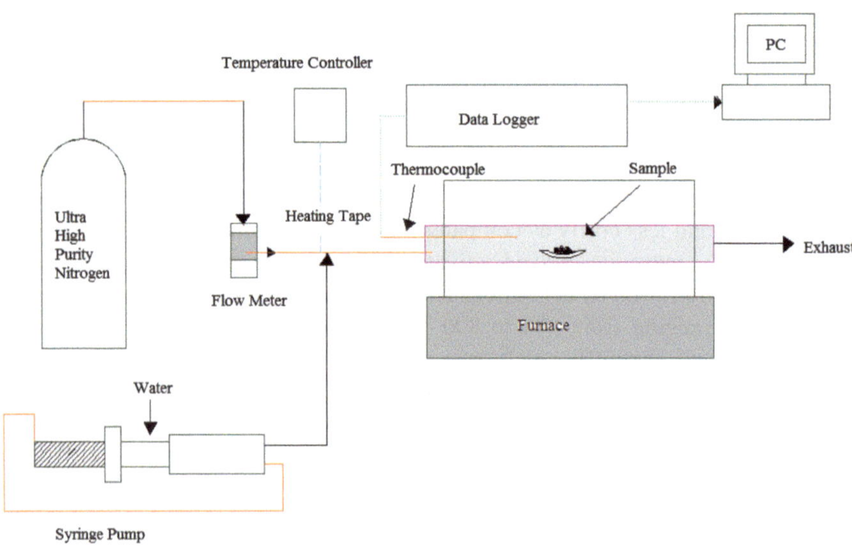

Fig. 6.1 Bench-scale experimental apparatus

the activating agent, the carbonization process, the chemical and physical activation procedures, and the characterization of the activated carbon samples descriptions are described below.

6.2.2 Sample Preparation

Mesquite tree bark was crushed and shredded using a blender or a chisel and a hammer to obtain a size of 2–5 mm. The sample was washed thrice with deionized water to remove the mud and dirt. The clean sample was filtered using a stainless steel strainer to remove water. The sample was then transferred into a tray to be dried in an oven (Sheldon Manufacturing, Model 1300U) at a temperature of 110 °C for 24 h to remove excess moisture content from the sample. Samples of 25 g were taken and labeled after the drying process.

6.2.3 Preparation of the Activating Agent

A solution of potassium carbonate (K_2CO_3) was prepared by mixing 100 g of chemical with 100 mL of deionized water in a standard beaker. The solution was mixed using a magnetic stirrer for 10 min to obtain a uniform and saturated solution of 7.2 M K_2CO_3.

6.2.4 Carbonization

The process of carbonization was carried out in a three-zone tubular-furnace (Lindberg/Blue, Model STF55346C-1). The dry wood sample was placed in a stainless steel mesh basket that was then placed inside a 3-in. diameter ceramic tube located in the three-zone temperature controller furnace. The wood sample was heated to temperatures ranging from 600 to 800 °C in UHP-N_2 (Matheson, 99.999 % pure) with a gas flow rate of 0.5 L/min controlled by a volumetric flow meter (Omega, FL-2060). The volumetric flow meter and gas flow rates were calibrated with a flow meter (BIOS, Drycal).

6.2.5 Chemical Activation Process

The mesquite wood samples were treated with the potassium carbonate solution for specified conditions of time and temperature. The treatment of wooden samples with chemical solutions is called impregnation. After chemical impregnation, the samples were washed out twice with deionized water and then soaked in a 0.5 N hydrochloric acid solution for 1 h to ensure neutralization of the sample. The samples were again washed with deionized water several times until the pH of the washed deionized water remained between 6 and 7. Finally, the washed samples were dried over night in the oven at 110 °C to remove all the moisture. Weights of the samples before and after the carbonization and activation treatments were measured using a scale (Scientech, S L 600 Rev-B).

6.2.6 *Physical Activation Process*

Samples of 25 g each were carbonized and physically activated simultaneously in the furnace with a gaseous mixture of steam and UHP-N_2. During the physical activation with steam, deionized water was injected at a flow rate of 0.37 mL/min using a syringe pump (New Era Pump Systems Inc., Model NE-300) to a flow of 0.5 L/min dry UHP-N_2. The temperature of the steam was controlled at 250 °C using heating tape and a step-less input temperature controller (Barnstead, Type 45500). Activation temperature and time conditions for the samples are provided in Table 6.1.

6.2.7 *Characterization of Wood-Derived Activated Carbon*

6.2.7.1 % Yield

The yield, an indicator of the efficiency of the activation process, is calculated from the resultant product weight and its initial sample weight as the ratio of final weight to initial weight. The samples were weighed on a scale. The weight of the dry sample before carbonization was taken as the initial weight and the weight of the dry sample after the carbonization and activation was taken as the final weight. The final weight was then divided by the initial weight. This was converted into a percentage.

6.2.7.2 Bulk Density

The bulk density includes both pore and inter-particle spaces. The weighed sample was transferred into a graduated cylinder to determine the volume. Dividing the sample weight by the total volume yields the bulk density (Ruthven, 1984).

Table 6.1 Process conditions and physical properties of WDAC samples

No.	Carbonization temperature (°C)	Activation time (min)	Bulk density (g/mL)	Yield (%)	N_2-BET surface area (m^2/g)	Total pore volume (cm^3/g)	Average pore width (Å)
Physical activation							
1	600	30	0.341	24.5	576	0.814	4.118
2	600	45	0.192	14.6	841	1.41	3.28
3	600	60	0.162	29.1	492	0.573	6.86
4	600	75	0.139	5.27	442	0.724	6.61
5	600	90	0.125	11.9	433	0.746	6.88
6	625	30	0.164	4.59	329	0.593	6.95
7	625	45	0.181	5.45	106	0.783	7.42
8	625	60	0.172	11.7	647	0.562	8.60
9	625	75	0.190	16.7	130	0.811	6.79
10	625	90	0.187	13.5	56	0.364	6.79
11	650	30	0.235	16.9	677	1.31	7.56
12	650	45	0.151	16.6	137	1.15	7.83
13	650	60	0.174	27.1	710	1.34	7.27
14	650	75	0.485	0.49	572	1.08	7.33
15	650	90	0.111	15.6	513	0.849	6.65
16	650	120	0.234	4.31	552	0.877	6.28
17	675	30	0.220	12.3	76	1.04	6.28
18	675	45	0.196	19.3	137	1.04	7.50
19	675	60	0.164	5.89	512	0.853	6.73
20	675	75	0.128	10.2	661	1.14	6.96
21	675	90	0.254	27.4	2	0.188	15.2
22	700	30	0.264	7.72	714	0.877	6.29
23	700	45	0.152	19.0	478	0.815	6.85
24	700	60	0.131	19.8	566	0.423	8.39
25	700	75	0.082	19.8	360	0.679	7.27
26	700	90	0.166	11.6	54	0.460	7.29
27	700	120	0.070	8.40	633	0.349	7.68
28	725	30	0.249	15.9	0.52	0.228	13.2
29	725	45	0.235	18.8	0.45	0.236	12.9
30	725	60	0.120	18.8	459	0.938	7.85
31	725	75	0.164	15.5	302	0.509	6.78
32	725	90	0.204	30.9	57.4	0.455	8.71
33	750	30	0.136	9.95	643	0.791	6.51
34	750	45	0.237	9.02	2.14	0.176	15.4
35	750	60	0.162	14.5	340	1.08	9.46
36	750	75	0.113	7.22	805	1.84	8.45
37	750	90	0.231	20.9	4.46	0.287	16.8
38	750	120	0.263	1.05	3.32	0.300	20.8
39	800	60	0.209	5.86	644	2.49	6.86
40	800	120	0.108	17.0	296	0.135	7.62
Chemical activation							
41	700	60	0.125	7.50	262	0.128	7.54
42	700	120	0.127	6.10	269	0.113	7.06

6.2.7.3 Nitrogen Adsorption Surface Area and Pore Size Distribution

N_2-adsorption isotherms for the wood-derived activated carbon samples were evaluated at 77 K using a high-speed surface area and pore size analyzer (Quantachrome, Nova 2200e Series). The samples were degassed in the analyzer for 2 h at 200 °C prior to the N_2 adsorption experiments. The resulting N_2 adsorption data was used to determine the BET-N_2 surface area, pore size distribution, total pore volume, and pore width. The pore size distribution was found using the micropore analysis (MP) method for the micropore region and the Barret–Joyner–Halenda (BJH) method for the meso- and macro-porous. Pore size distribution determines the fraction of the total pore volume that is available to molecules of a certain size and shape.

6.2.7.4 Methanol and Water Vapor Equilibrium Adsorption Capacity

The amount of adsorbate that adsorbs onto any adsorbent is a function of concentration, temperature, and the properties of the adsorbent and adsorbate. A symmetrical gravimetric organic vapor sorption analyzer (VTI Inc., Model SGA100) was used to determine the equilibrium adsorption capacities of methanol and water vapor using the activated carbon samples. The gas stream loaded with the adsorbate was passed through a saturator/humidifier. UHP-N_2 was utilized as the carrier gas in the methanol and water vapor adsorption experiments. The relative humidity for the water vapor adsorption experiments was controlled with a dew point analyzer. Equilibrium adsorption isotherms were measured for the adsorbate's relative pressures ranging from 0 to 0.9 at ambient temperature.

6.2.8 *Adsorption Isotherm Modeling*

The Freundlich and Langmuir adsorption equations were used in this study to model the experimental data. The Freundlich model is an empirical expression which is represented by a power law equation, a commonly used model for a single solute system. In this model, the adsorption capacity is linear for a change in adsorbate concentration when plotted with log-log scales. The Freundlich equation is represented by (Ruthven, 1984):

$$q = KC^{\frac{1}{n}} \tag{6.1}$$

where q is the volume of adsorbate adsorbed per gram of adsorbent, C is the concentrate of the adsorbate, and K and n are constants. Values of $n > 1$ imply that affinity for the adsorbate decreases with increasing adsorption density. K and n

are determined from a plot of the log (q) against log C, where K is the intercept and $1/n$ is the slope.

The Langmuir isotherm model assumes monolayer adsorption on a homogeneous surface. This equation is expressed as follows:

$$q = \frac{q_m bC}{1 + C} \tag{6.2}$$

where q_m is the maximum capacity of adsorbent for adsorbate, b is the Langmuir empirical constant. The Langmuir adsorption parameters q_m and b can be determined with experimental adsorption data.

6.3 Results and Discussion

Carbonization and activation process conditions, the bulk density, percent yield, N_2-BET surface area, total pore volume, and average pore width of the resultant wood-derived activated carbon (WDAC) samples are provided in Table 6.1. Physically activated WDAC samples 1–40 were pyrolysed from 500 to 800 °C for times ranging from 30 to 120 min. Chemically activated WDAC samples 41 and 42 were pyrolysed at 700 °C at 60 and 120 min. These chemically activated samples were prepared for comparison purposes to the physical activation of the mesquite samples. The sample percent yield decreased with the extent of activation, and the lowest yields were from the chemically activated samples. The yield ranged from 0.49 to 30.90 % when the samples were activated with steam, while the percent yield was less than 7.50 % when the samples were treated with potassium carbonate solution. The samples that underwent physical activation with steam had a bulk density ranging from 0.070 to 0.485 g/mL which decreased with increases in the carbonization temperature. The bulk density for the chemically activated samples was not affected by the extent of activation, with an average bulk density value of 0.126 g/mL. The N_2-BET surface area, total pore volume, and average pore width for all the WDAC samples ranged from 0.450 to 841 m^2/g, from 0.135 to 2.489 cm^3/g, and from 3.3 to 20.8 Å, respectively. The WDAC sample with a carbonization temperature of 600 °C and 45 min of activation time had the highest N_2-BET surface area.

Physical properties of activated carbon fiber cloth (ACFC), granular activated carbon (GAC), and activated carbon monolith (ACM) are provided for comparison in Table 6.2. In general, ACFC, GAC, and ACM show larger values of bulk density, BET-N_2 surface area, and total pore volume when compared to the WDAC samples.

The micro and mesopore size distributions for selected WDAC samples are provided in Fig. 6.2. WDAC-1, WDAC-2, and WDAC-3 are the physically activated samples No. 24, 27, and 40, respectively from Table 6.1. WDAC-4 and WDAC-5 are the chemically activated samples No. 41 and 42, respectively from

Table 6.2 Physical properties of ACFC, GAC and ACM (Luo et al., 2006)

Sample	Bulk density (g/cm^3)	N_2 BET-surface area (m^2/g)	Total pore volume (cm^3/g)	Average pore width (Å)
ACFC[a]	2.03	1,604	0.803	7.40
GAC[b]	0.45	965	0.615	25.5
ACM[c]	1.04	603	0.800	15.2

[a]*ACFC* Activated carbon fiber cloth (ACFC-5092-20, American Kynol, Inc.)
[b]*GAC* Granular activated carbon (Carbon BPL)
[c]*ACM* Activated carbon monolith (RIDC, Beijing)

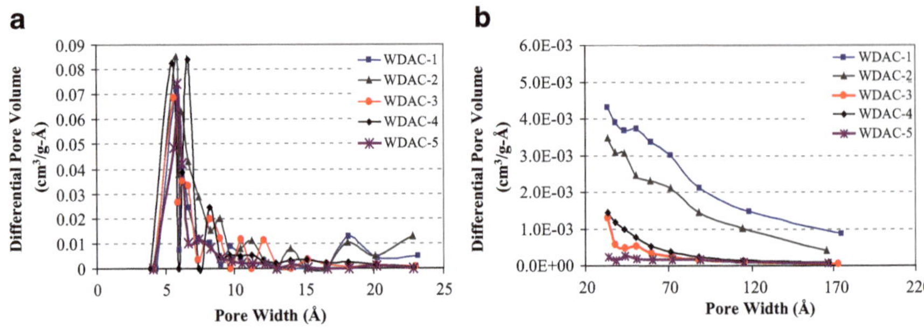

Fig. 6.2 Pore size distribution for WDAC samples using (**a**) the MP method (Micropore region) and (**b**) the BJH method (Mesopore region)

Table 6.1. The WDAC-1-3 produced by the physical method showed larger mesopores as compared to the ones that were chemically activated, while all WDAC samples showed similar differential pore volume in the micropore region. The pore width of these WDAC samples in the micropore region showed predominant values between 6 and 8 Å. The specific surface area of the samples decreased as the activation time and bulk density increased. An extended time of activation caused the mesquite sample to lose significant amount of bulk material causing a reduction in the specific surface area of the sample.

Plots of the N_2-BET-surface area as a function of the activation time, bulk density, and carbonization time for selected WDAC samples are provided in Figs. 6.3, 6.4, and 6.5. An increase in the activation time resulted in a lower surface area with more mass of carbon lost that resulted in a smaller bulk density of the carbonaceous material.

6.3.1 Equilibrium Adsorption Capacity

6.3.1.1 Water Vapor Adsorption Results

High humidity in the pollutant gas streams affects the adsorption capacity of activated carbon (Sullivan et al., 2007). The water hysteresis effect on adsorption

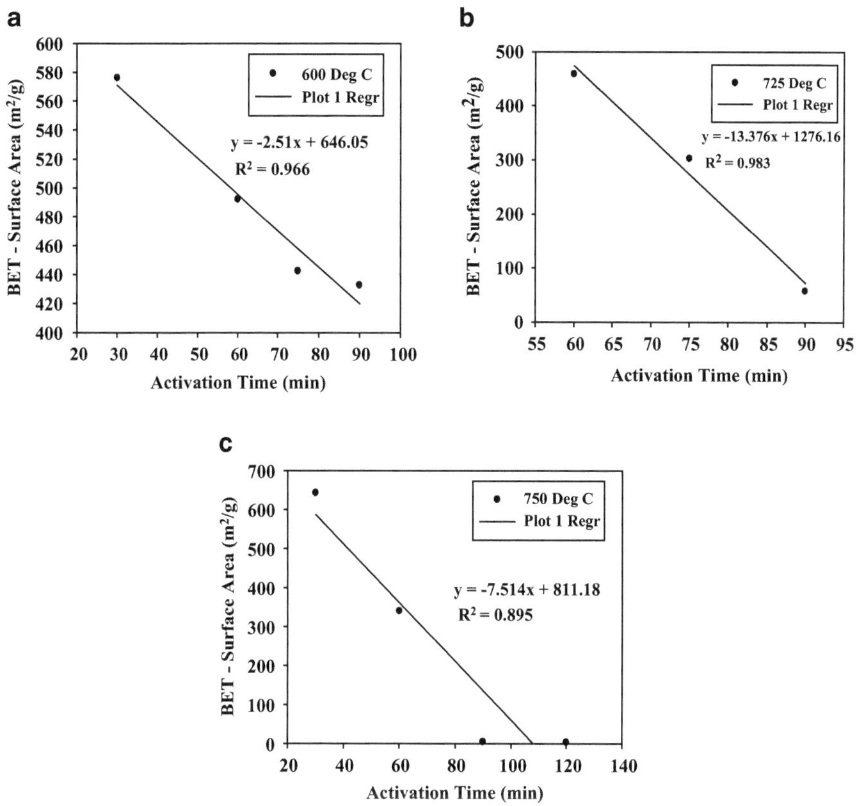

Fig. 6.3 N_2-BET surface area of WDAC samples as a function of activation time at (**a**) 600 °C, (**b**) 725 °C and (**c**) 750 °C

is observed by adsorbing water vapor on the WDAC samples at different water vapor relative pressures. Plots of water vapor adsorption capacity as a function of relative humidity at 25 °C are shown in Fig. 6.6. The adsorption capacities increased with increasing water vapor relative humidity. This suggests an adsorption competition of water vapor and pollutants for adsorption sites in gas streams with high humidities. It was observed that the hysteresis effect for WDAC samples was less pronounced than for ACFC, ACM, and GAC during physisorption. For the selected WDAC samples the adsorption capacities of water vapor at relative humidity of 90 % were 1.0, 0.93, 0.7 $g_{water}/g_{adsorbent}$, respectively, whereas for ACFC, GAC, and ACM the water vapor adsorption capacities were 0.66, 0.37 and 0.25 $g_{water}/g_{adsorbent}$, respectively. The selected WDAC samples which underwent chemical activation had an adsorption capacity of 0.12 $g_{water}/g_{adsorbent}$ at a relative humidity of 90 % (Fig. 6.6b). The physically activated WDAC samples had higher water vapor adsorption capacities than the other samples. Figure 6.7 compares the water vapor adsorption capacities for all studied adsorbents.

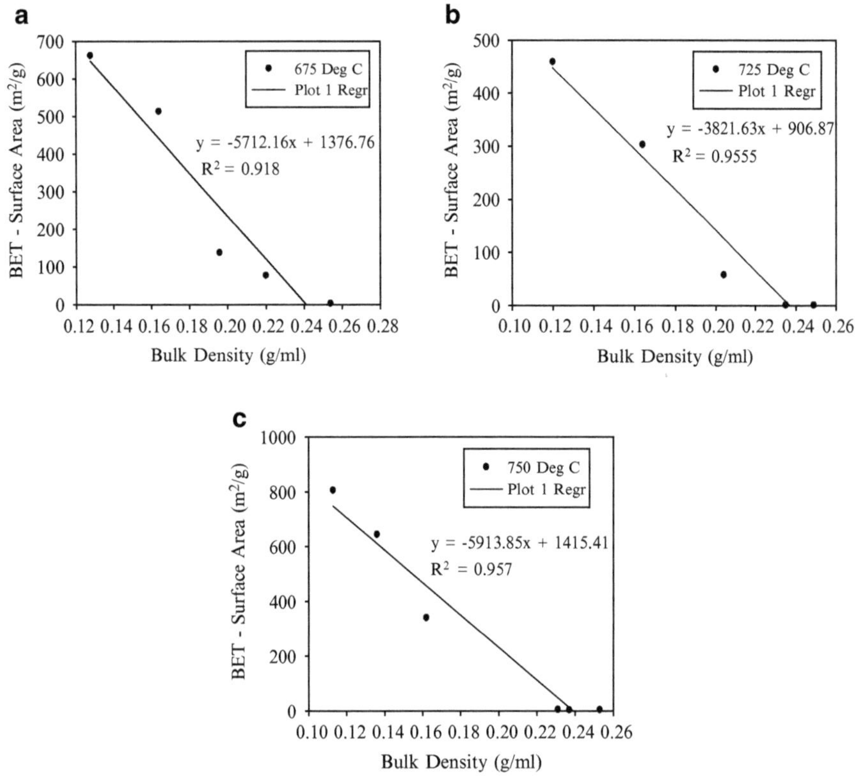

Fig. 6.4 N_2-BET surface area of WDAC samples as a function of bulk density at (**a**) 675 °C, (**b**) 725 °C and (**c**) 750 °C

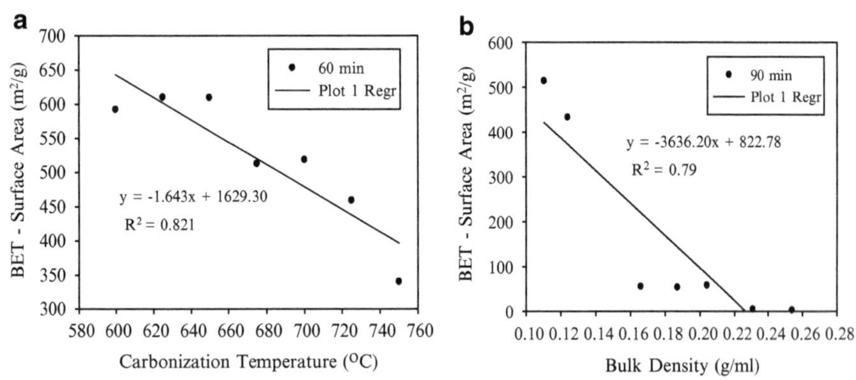

Fig. 6.5 N_2-BET surface area of WDAC samples (**a**) as a function of carbonization temperature at 60 min of carbonization time, and (**b**) as a function of bulk density at 90 min of carbonization time

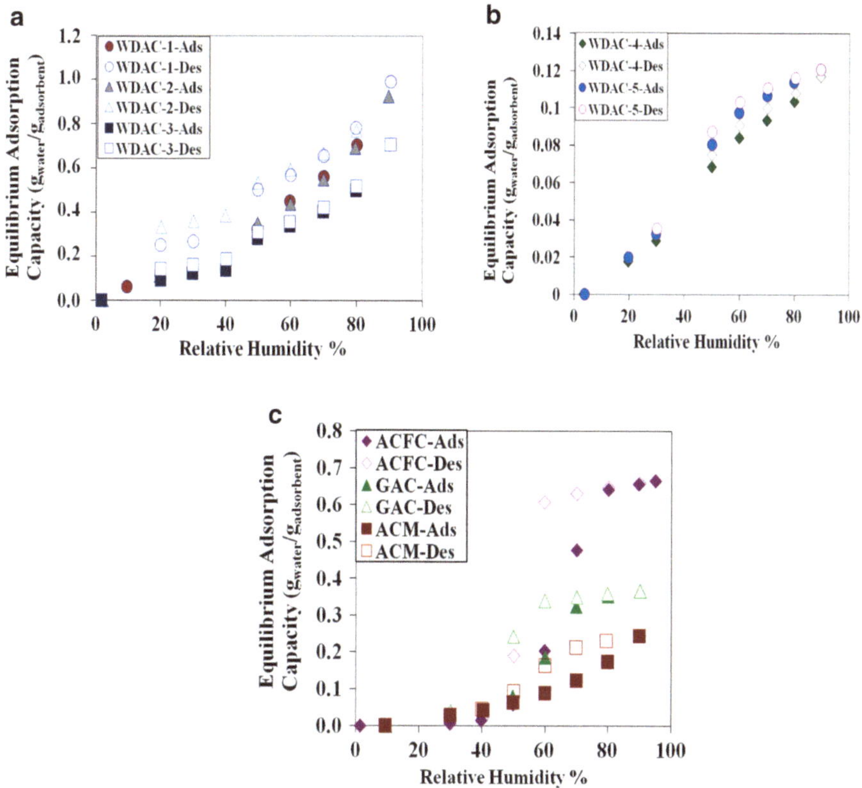

Fig. 6.6 Equilibrium adsorption and desorption capacities of water vapor onto (**a**) selected physically activated WDAC samples, (**b**) the chemically activated WDAC samples, (**c**) commercially available adsorbents: ACFC, ACM, and GAC

Fig. 6.7 Comparison of adsorption capacities for all the samples

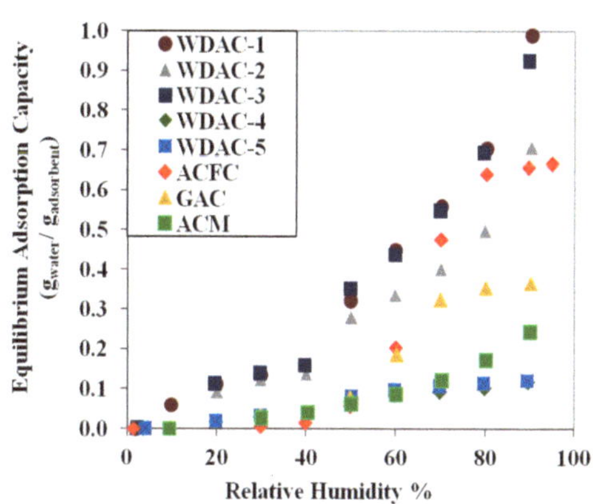

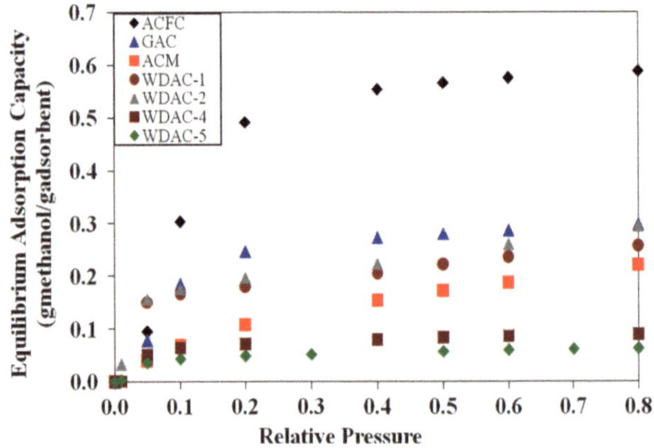

Fig. 6.8 Comparison of the equilibrium adsorption capacities of methanol using selected WDAC, ACFC, GAC, and ACM samples

6.3.1.2 Methanol Adsorption Results

A plot of the methanol adsorption capacity as a function of methanol relative pressure for selected WDAC samples, ACFC, ACM, and GAC is provided in Fig. 6.8. Methanol was used because it is an organic solvent widely used in industry. Methanol is classified as hazardous air pollutant (HAP) and is listed in the top 3 of HAPs emitted to the atmosphere in 2008 (United States Environmental Protection Agency [U.S. EPA], 2010). The adsorption capacities of methanol at a relative pressure of 0.1 for WDAC-1, WDAC-2, WDAC-4, WDAC-5, ACFC, GAC, and ACM were 0.17, 0.18, 0.06, 0.04, 0.30, 0.18, and 0.07 $g_{methanol}/g_{adsorbent}$, respectively. For the chemically activated WDAC samples (WDAC-4 and -5), a minimum increase in the adsorption capacity of methanol from 0.04 to 0.06 $g_{methanol}/g_{adsorbent}$ was observed when the methanol relative pressure increased from 0.1 to 0.9. At relative pressures <0.1, which are typical concentrations in industrial emissions, the physically activated WDAC samples had the highest adsorption capacities of methanol as compared to the commercially available ACFC, GAC, and ACM adsorbents. More studies are needed to better understand the adsorption mechanism occurring at low relative pressures.

Plots of experimental, Freundlich, and Langmuir model adsorption isotherms at 25 °C are shown in Figs. 6.9 and 6.10. The adsorption capacities increased with increasing adsorbate relative pressure for both the experimental and model results. There was a correlation among the experimental data results, the Freundlich, and the Langmuir models for each of the samples. Adsorption isotherms followed a type I isotherm according to the Brunauer classification. Table 6.3 provides the parameters of the Freundlich and Langmuir equations fitted to the experimental adsorption data. The values for the Freundlich parameter K were 23.4, 1.97, and 0.781 for ACFC, GAC, and ACM, respectively, while the values for WDAC-1, WDAC-2, WDAC-4, and WDAC-5 samples were 32.0, 19.4, 8.96, and 6.19, respectively. Values for $n > 1$ indicated that the affinity for the adsorbate decreased with increasing adsorption density.

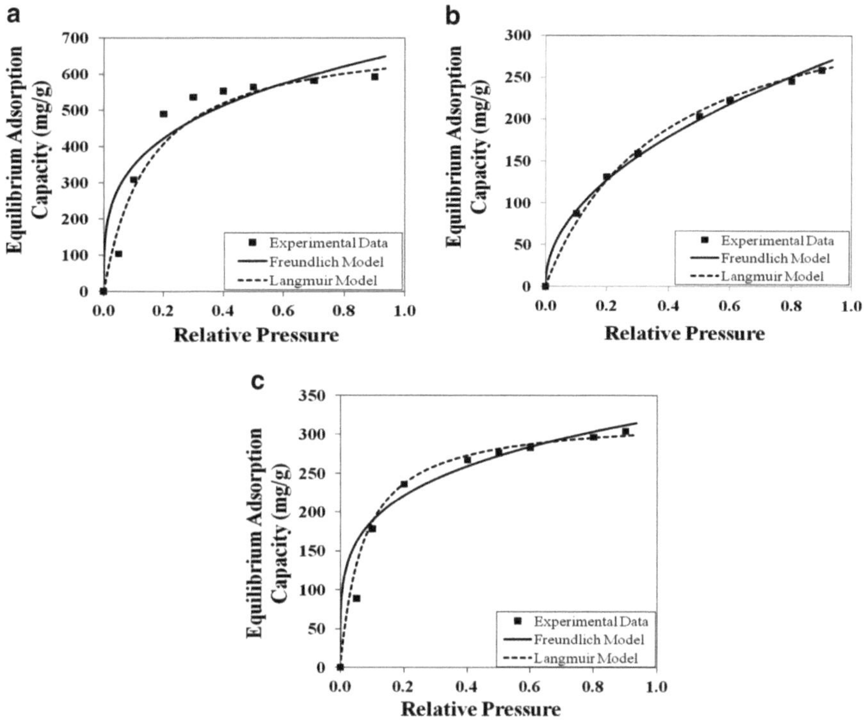

Fig. 6.9 Comparison of Freundlich and Langmuir adsorption isotherms with experimental data – equilibrium adsorption capacities of methanol onto (**a**) ACFC, (**b**) ACM and (**c**) GAC

6.4 Summary and Conclusions

The wood-derived activated carbons (WDACs) produced from mesquite wood materials have the potential to adsorb gaseous compounds with high adsorption capacities at low relative pressures which are the prevalent industrial conditions. The proposed steam activation method to manufacture WDAC creates a recycle and re-use path for the waste mesquite wood materials in South Texas. The manufacture of WDAC through chemical activation is not environmentally friendly because of the disposal of the hazardous chemicals used in the activation process.

In this study, the chemical activation was achieved using potassium carbonate as the activating reagent, whereas the physical activation used steam as the activating agent. Effects of activation time and carbonization temperatures on the N_2-BET surface areas of activated carbon samples were studied. Samples were characterized by N_2-BET surface area, pore volume, pore size distribution, bulk density, percent yield, and equilibrium adsorption capacities for water vapor and methanol. For a constant carbonization temperature, the pore volume and pore width of the activated carbons are enhanced with an increased activation time. For a constant

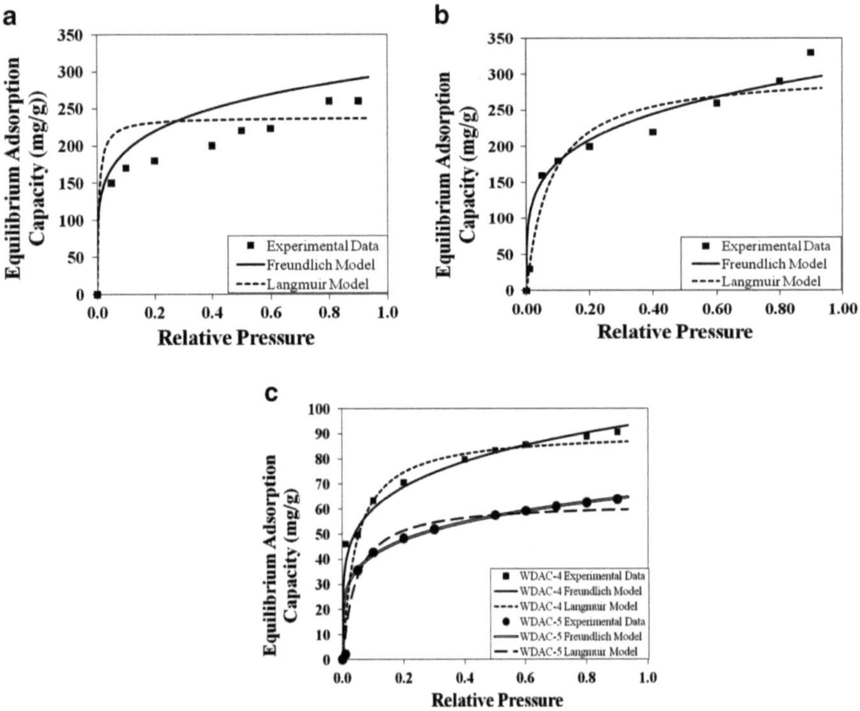

Fig. 6.10 Comparison of Freundlich and Langmuir adsorption isotherms with experimental data – equilibrium adsorption capacities of methanol onto (**a**) WDAC-1, (**b**) WDAC-2 (**c**) WDAC-4 and WDAC-5 samples

Table 6.3 Freundlich and Langmuir parameters for commercially available and WDAC samples at 25 °C

	Freundlich	Model	Langumuir model	
Sample	K	N	q_m	B
ACFC	23.4	3.59	714	0.00004
GAC	1.97	2.69	323	0.00008
ACM	0.781	2.04	370	0.00002
WDAC-1	32.0	5.40	238	0.00098
WDAC-2	19.4	.0229	303	0.00008
WDAC-4	8.96	5.10	90.9	0.00001
WDAC-5	6.19	5.09	62.5	0.00014

activation time of 60 min, a maximum surface area was achieved when the activation temperature was 650 °C. WDAC surface areas of 633, 841, and 805 m^2/g were achieved when the operational conditions were 700 °C for 120 min, 600 °C for 45 min, and 750 °C for 75 min, respectively. The physically activated WDAC samples proved similar or superior adsorption capacities of methanol compared to commercial activated carbon at low relative pressures. WDAC samples were more hydrophilic because they adsorbed more water vapor than commercial activated carbon materials.

Acknowledgments This material is based upon work supported by the Center of Research Excellence in Science and Technology – Research on Environmental Sustainability of Semi-Arid Coastal Areas (CREST-RESSACA) at Texas A&M University-Kingsville (TAMUK) through a Cooperative Agreement (No. HRD-0734850) from the National Science Foundation (NSF) and Texas A&M University-Kingsville. Any opinions, findings, and conclusions or recommendations expressed in this material are those of the authors and do not necessarily reflect the views of the National Science Foundation or Texas A&M University-Kingsville.

References

Adinata, D., Daud, W., & Aroua, K. (2005). Preparation and characterization of activated carbon from palm shell by chemical activation with K_2CO_3. *Bioresource Technology, 98*, 145–149.

Avelar, F., Bianchi, M., Goncalves, M., & Gaspar da Mota, E. (2010). The use piassava fibers (Attalea funifera) in the preparation of activated carbon. *Bioresource Technology, 101*, 4639–4645.

Baccara, R., Bouzida, J., Fekib, M., & Montiela, A. (2009). Preparation of activated carbon from Tunisian olive-waste cakes and its application for adsorption of heavy metal ions. *Journal of Hazardous Materials, 162*, 1522–1529.

Bansal, R., Donnet, J., Stoeckli, F., & Dekker, M. (1988). *Active carbon*. New York: Marcel Dekker.

Bansode, R., Losso, J., Marshall, W., Rao, R., & Portier, R. (2004). Pecan shell-based granular activated carbon for treatment of chemical oxygen demand (COD) in municipal wastewater. *Bioresource Technology, 94*, 129–135.

Hayashi, J., Kazehayaa, A., Muroyama, K., & Watkinson, A. (2000). Preparation of activated carbon from lignin by chemical activation. *Carbon, 38*, 1873–1878.

Kantarli, I., & Yanik, J. (2010). Activated carbon from leather shaving wastes and its application in removal of toxic materials. *Journal of Hazardous Materials, 179*, 348–356.

Khezami, L., Chetouani, A., Taouk, R., & Capart, R. (2005). Production and characterization of activated carbon from wood components in powder: Cellulose, lignin, xylan. *Powder Technology, 157*, 48–56.

Lehmann, C., Rostam-Abadi, M., Rood, M., & Sun, J. (1998). Reprocessing and reuse of waste tire rubber to solve air-quality related problems. *Energy & Fuels, 12*, 1095–1099.

Lima, I., & Marshall, W. (2005). Adsorption of selected environmentally important metals by poultry manure-based granular activated carbon. *Journal of Chemical Technology and Biotechnology, 80*, 1054–1061.

Lin, Y., & Teng, H. (2003). A novel method for carbon modification with minute polyaniline deposition to enhance the capacitance of porous carbon electrodes. *Carbon, 41*, 2865–2871.

Lua, A., & Yang, T. (2004). Effects of vacuum pyrolysis conditions on the characteristics of activated carbons derived from pistachio-nut shells. *Colloid and Interface Science, 276*, 364–372.

Luo, L., Ramirez, D., Rood, M., Grevillot, G., Hay, K., & Thurston, D. (2006). Adsorption and electrothermal desorption of organic vapors using activated carbon adsorbents with novel morphologies. *Carbon, 44*, 2715–2723.

Qui, G., & Guo, M. (2009). Quality of poultry litter-derived granular activated carbon. *Bioresource Technology, 101*, 379–386.

Ramirez, D., Sullivan, P., Rood, M., & Hay, K. (2004). Equilibrium adsorption of phenol-, tire-, and coal-derived activated carbons for organic vapors. *Journal of Environmental Engineering, ASCE, 130*, 231–241.

Rosas, M., Bedia, J., Mirasol, J., & Cordero, T. (2008). Preparation of hemp-derived activated carbon monoliths. Adsorption of water vapor. *Industrial and Engineering Chemistry Research, 47*, 1288–1296.

Ruthven, D. (1984). *Principles of adsorption and adsorption processes*. New York: Wiley.
Sudaryanto, Y., Hartono, S., Irawaty, W., Hindarso, H., & Ismadji, S. (2006). High surface area activated carbon prepared from cassava peel by chemical activation. *Bioresource Technology, 97*, 734–739.
Suhas, P., Carrott, M., & Carrott, R. (2007). Lignin – From natural adsorbent to activated carbon: A review. *Bioresource Technology, 98*, 2301–2312.
Sullivan, P., Rood, M., Dombrowski, K., & Hay, K. (2004). Capture of organic vapors using adsorption and electrothermal regeneration. *Journal of Environmental Engineering, 130*, 258–267.
Sullivan, P., Stone, B., Hashisho, Z., & Rood, M. (2007). Water adsorption with hysteresis effect onto microporous activated carbon fabrics. *Adsorption, 13*, 173–189.
United States Environmental Protection Agency [U.S. EPA]. (2010). *2008 National emission inventory, air pollutant emission trends*. Retrieved September 1, 2010, from http://www.epa.gov/ttn/chief/trends/index.html
Wua, F., Tsengb, R., Hu, C., & Wang, C. (2004). Physical and electrochemical characterization of activated carbons prepared from fir woods for super capacitors. *Journal of Power Sources, 138*, 351–359.
Yanhe, H., Xie, Q., Shuo, C., Shibo, W., & Yaobin, Z. (2007). Electrochemical enhancement of adsorption capacity of activated carbon fibers and their surface physicochemical characterizations. *Electrochimica Acta, 52*, 3075–3081.

Chapter 7
Assessment of Nutrient and Heavy Metal Levels in Manadas Creek, an Urban Tributary of the Rio Grande in Laredo, Texas

Brianna Flores, Jianhong Ren, Sushma Krishnamurthy, and Wayne Belzer

Abstract For many years the pollution of the Rio Grande has led to the extinction or reduction of certain wildlife species and affected the health of the residents along the border. Although great strides have been made in monitoring the Rio Grande, there has been a lack of intensive studies on the portion of the river located in South Texas. This paper presents an initial water and sediment quality study conducted in 2006 at a site next to a decommissioned antimony plant on Manadas Creek, a tributary of the Rio Grande located in Laredo, Texas. Results show that Manadas Creek had elevated concentrations of antimony (Sb) and arsenic (As) in June through December, total phosphorous in August and September, and ortho-phosphorous in August. All nutrients and dissolved oxygen showed significant temporal variations; however, for metals, only the nickel (Ni) and cobalt (Co) detected in the water phase had a significant temporal variation among the sampling periods. Strong correlations were found between the stream flow and sediments associated As, Sb, and chromium (Cr), among the trace metal elements

B. Flores
Department of Environmental Engineering, Frank H. Dotterweich College of Engineering, Texas A&M University-Kingsville, Kingsville, TX 78363-8202, USA

Surface Mining and Reclamation Division, Railroad Commission of Texas, Austin, TX, USA

J. Ren (✉)
Department of Environmental Engineering, Frank H. Dotterweich College of Engineering, Texas A&M University-Kingsville, Kingsville, TX 78363-8202, USA
e-mail: jianhong.ren@tamuk.edu

S. Krishnamurthy
School of Sciences, University of Louisiana at Monroe, Monroe, LA 71209, USA

W. Belzer
Environmental Management Division, International Boundary and Water Commission, United States Section, Clean Rivers Program, El Paso, TX, USA

D. Ramirez et al. (eds.), *Environmental Sustainability Issues in the South Texas–Mexico Border Region*, DOI 10.1007/978-94-007-7122-2_7,

detected, and among the nutrients detected. These results contribute to a better understanding of potential environmental problems associated with water quality of the Rio Grande in the Laredo area.

Keywords Nutrients • Metals • Rio Grande • Manadas Creek • Water and sediments

7.1 Introduction

The Rio Grande stretches over 2,000 miles from the southern Rocky Mountains in Colorado to the tip of Texas where it meets the Gulf of Mexico. It is the natural boundary between the U.S. and Mexico from just upstream of El Paso, Texas, to below Brownsville, Texas. The Rio Grande supports rapidly growing cities in three states in the United States and four in Mexico. The stretch of the river from Laredo/Nuevo Laredo to the mouth of the Rio Grande constitutes the primary drinking water source for the majority of the population in both countries (United States Section International Boundary and Water Commission [U.S.IBWC], 1994). In addition, the river supports agriculture and recreation for thousands of people on both sides of the border. Wildlife also heavily relies upon the river for survival.

The City of Laredo is one of the most rapidly growing cities in the United States. It is a major port of entry for international trade and tourism between the United States and Mexico. A wide variety of products, such as steel, clothing, candles, military supplies, and leather goods, are manufactured in the area. With an increase in factories due to the passage of North America Free Trade Agreement (NAFTA) in the early 1990s and increased trade with Mexico, the population in the area has increased significantly, which has led to many infrastructure changes and an increase in pollution. Although numerous studies examining the water quality of the Rio Grande have been conducted by the Texas Commission on Environmental Quality (TCEQ) and the United States Section International Boundary and Water Commission (U.S.IBWC) (Texas Natural Resource Conservation Commission [TNRCC], 2002; U.S.IBWC, 2003), more intensive studies of the Rio Grande water quality in the Laredo area are still lacking.

Manadas Creek is a tributary of the Rio Grande. The creek's watershed is located in a highly industrial area of Laredo where manufacturing facilities, trucking, and warehouses exist (TNRCC, 2002). In 1994 and 1997, the TCEQ and the U.S.IBWC found elevated levels of chromium (Cr), mercury (Hg), nickel (Ni), thallium (Tl), 4,4′-dichlorodiphenyldichloroethylene (4,4′-DDE), chloride, Sb, As, and dichloro-diphenyl-trichloroethane (DDT) in the water of Manadas Creek. The site sampled in this binational study was located only 0.8 km upstream from the confluence of the creek and the Rio Grande.

The environmental and health impacts of heavy metals and nutrients have been documented by many researchers. Mendoza et al. (2004) indicated that metals may be associated with deformations in aquatic life and leukemia and congenital malformation in humans. Mercury (Hg) in the form of methylmercury can impair neurological development in fetuses, infants, and children (U.S. Environmental Protection Agency [U.S. EPA], 2006). Chronic inhalation exposure to Ni in humans can result in respiratory effects including a type of asthma specific to Ni, decreased lung function, and bronchitis (Agency for Toxic Substances, Disease Registry [ATSDR], 2005). Excessive levels of nutrients such as nitrate (NO_3^-), nitrite (NO_2^-), total phosphorus (TP), and ortho-phosphorus (OP) can lead to eutrophication. Eutrophication causes a variety of negative impacts on the water quality and ecosystem health including depleted dissolved oxygen (DO) due to accelerated production of organic matter particularly algae, nuisance and toxic algal blooms, and loss of submerged aquatic vegetation (Bricker et al., 1999; TNRCC, 2002).

In a previous paper (Flores et al., 2009), results on organochlorine pesticides obtained from July through December 2006 were reported at the same sampling site used in this study. A study, which was based on the findings to be presented in this paper and focused on spatial distribution of arsenic and antimony in Manadas Creek in 2008 was also published by the same research group (Baeza et al., 2010). The present paper will document preliminary results of nutrients, heavy metals, and conventional parameters including temperature, pH, DO, salinity, and total dissolved solids (TDS) collected in 2006. It will provide a preliminary perspective of the temporal variation of nutrients and heavy metals at the study site. These results will contribute to filling in the gaps in the Rio Grande water quality database and provide further evidence of the contamination levels of the Rio Grande in the Laredo area.

7.2 Research Methods

7.2.1 Description of the Study Site

Manadas Creek receives waters originating from the surrounding creek watershed with a total drainage area of 40.96 km^2. The study site located in Manadas Creek is 1.57 km from the confluence of the creek and the Rio Grande within the Laredo city limits. The site has warehouses, a major railway line, and Interstate-35 in its vicinity. In particular, the sampling site lies adjacent to a 102-acre site housing a decommissioned antimony plant. The plant was established in 1928 and housed a blast furnace until 1992 to process antimony-bearing ore to produce antimony oxide and other antimony compounds. Two slag sites with antimony byproducts still remain at the site for remediation resulting in the raising of numerous public concerns (TNRCC, 2002). In addition, a wastewater treatment plant (WWTP) is located 5.64 km upstream of the study site. A portion of the WWTP effluent is released into Manadas Creek.

Two sub-locations within this study site, MC-1 (27°34′31.02″ N and 99°30′21.90″ W) and MC-2 (27°34′31.02″ N and 99° 30′ 21.78″ W), were sampled

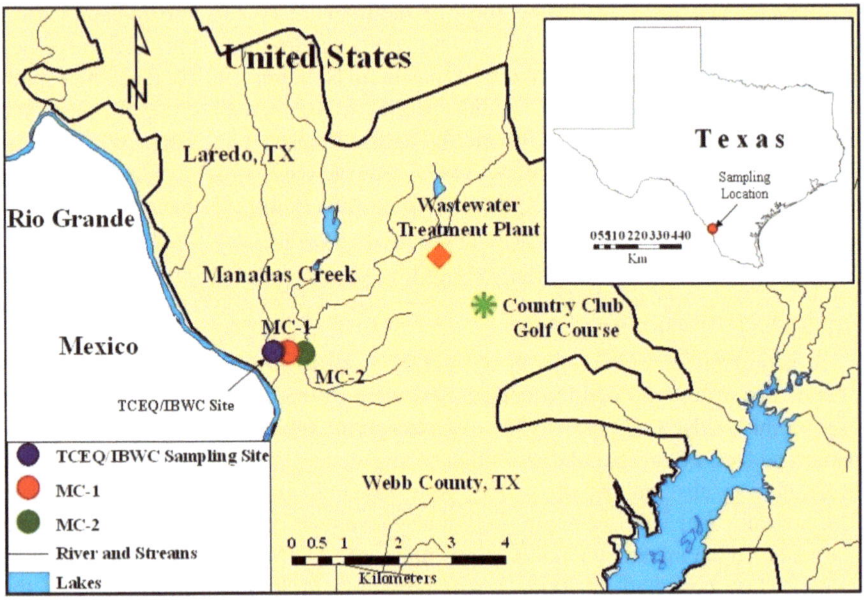

Fig. 7.1 Map of sampling sites

in 2006. These sub-locations are 11.4 m apart (Fig. 7.1). They were chosen based on accessibility, proximity to the Rio Grande, and location within the Laredo area. Since the beginning of April 2006, this site has undergone dramatic changes with the construction of a bridge over the creek. Most of the brush and vegetation along the banks of the creek adjacent to the bridge were removed and replaced by concrete.

7.2.2 Water and Sediment Sample Collection

Samples were collected on 11 sampling dates – January 31, February 28, March 28, April 27, June 9, July 28, August 25, September 22, October 27, November 27, and December 18, 2006. A total of 22 water samples were collected on the sampling dates of January through December of 2006 for nutrients analysis. A total of 12 water samples and 12 sediment samples were collected on the sampling dates of July through December of 2006 for metal analysis. All samples were collected at approximately 11:00 a.m. on each of the sampling dates. Field parameters including pH, DO, temperature, salinity, and conductivity were measured in-situ using an YSI 556 multi-probe (YSI Inc., Yellow Springs, OH). Rainfall information was downloaded from the Texas Weather Connection website (http://twc.tamu.edu/). The stream flow was measured using a FlowTracker (SonTek/YSI, San Diego, CA) during each sampling trip.

Water samples were collected following the TCEQ procedures of submerging pre-cleaned and closed amber 1-L plastic jars into the water at 6″ below the water surface (Texas Commission on Environmental Quality [TCEQ], 2003). The jar caps were slowly opened under water to allow the jars to completely fill with water. The jars were then closed while still submerged and pulled out of the water. The collected water samples were stored in 1-L, amber, wide-mouth, plastic bottles, and preserved with 1–2 mL H_2SO_4 for the analysis of NO_3^- and total phosphorous (TP). Amber, 1-L, glass jars were used for the collection and storage of water samples for analysis of NO_2^-. These samples were analyzed within 48 h after collection (HACH, 2003). Water samples collected to determine ortho-phosphorus (OP) levels were filtered on site using 0.45 μm filters and analyzed within 48 h after collection. The water samples collected for metal analysis were filtered using 0.45 μm filters, preserved with HNO_3, and stored in 1-L, amber, wide-mouth, plastic bottles. All collected samples were placed on ice, transported back to the laboratory immediately, and stored at 4 °C until further analysis.

Sediment samples were collected at the same location as the water samples following the TCEQ procedures (TCEQ, 2003). A stainless steel garden trowel was used to gather the surface sediments. The trowel was lowered 1–1.5 ft to the bottom of the creek, with special care being taken to avoid disturbing any sediment. Small rocks and pebbles in the first inch of the sediment were scraped off using the tip of the trowel and discarded. A 6–8 in. layer of surficial fine sediments were then collected and placed in pre-cleaned 500 mL, short amber glass wide-mouth jars with Teflon lined lids and transported back to the laboratory on ice for further analysis.

7.2.3 Analysis of Water Quality Parameters

The parameters measured in the laboratory included OP, TP, NO_2^-, NO_3^-, and heavy metals. All nutrient analyses were conducted within 48 h of collection (HACH, 2003). Nutrients were analyzed using a HACH 2500 spectrophotometer. NO_3^- was analyzed using the Cadmium Reduction Method (HACH 8039), NO_2^- by the HACH 8153 method, TP by the HACH 8190 method, and OP by the HACH 8048 method. The NO_2^- standard was purchased from Ricca Chemical Company, Arlington, Texas, and the NO_3^- and P standards were from HACH Company World Headquarters, Loveland, Colorado.

Water and sediment samples were analyzed within 3 months of collection for heavy metals. For the determination of dissolved metal concentrations, Inductively Coupled Plasma-Mass Spectrometry (ICP-MS) (Thermo Electron Corporation, Waltham, MA) was used following the procedures described in *Standard Methods* (American Public Health Association [APHA], 1992) in conjunction with U.S.EPA method 208.1. Standard Tune D solution consisting of 10 mg/L of 24 metal elements (Al, As, Ba, Be, Ca, Cd, Co, Cr, Cu, Fe, In, K, Li, Mg, Mn, Mo, Na, Ni, Pb, Se, Tl, V, U, and Zn) in 5 % HNO_3, respectively, was purchased from Thermo Electron Corporation (Waltham, MA). The Sb standard was purchased from Inorganic Ventures (Lakewood, NJ) with a concentration of 100 ± 0.037 mg/L in 0.7 % HNO_3.

For determination of the bioavailable fraction of metals associated with sediments, the sediment samples were extracted using a sequential extraction method of

Silveira et al. (2006) before analysis using the ICP-MS. Sequential extractions were carried out in triplicate using 1 g of air-dried sediment. In detail, sediment samples were placed in 50-mL polycarbonate centrifuge tubes and mixed in a stepwise fashion with various reagents (15 mL 0.1 M $CaCl_2$ for soluble-exchangeable metals, 30 mL 1.0 M NaOAC for surface adsorbed metals, and 5 mL NaOCl for organic metals). The suspensions were then equilibrated as suggested by the method (2 h for soluble-exchangeable metal extraction, 5 h for surface adsorbed metal extraction, and 30 min for organic metal extraction). Following each equilibration, the solution and solid phases were separated by centrifugation at 2,900 rpm (1,225 g) for 10 min. The solid residues were then resuspended in 5 mL of 0.1 M NaCl, shaken by hand, and centrifuged to displace the extracting solution remaining from the previous step between each successive extraction. The supernatant liquid was then added to the former extractant. Once the sequential extraction was complete, the extractant was filtered using 0.45 μm filters, preserved using 2 % HNO_3, and analyzed using an ICP-MS.

7.2.4 Quality Control

Several types of quality control measures were employed during the course of this study to ensure the validity of the data reported. All sampling containers used to collect and store water and sediment samples were filled with 10 % HNO_3 a week prior to use. Before use in the field, the containers were rinsed with deionized water in the laboratory followed by a thorough rinse with the stream water collected at the sampling site in the field. All sampling personnel obtained training from U.S.IBWC on quality control precautions for effective field sampling. Each sample analysis was preformed three times to ensure the reliability of the measurements. Individual calibration curves were preformed for each nutrient and metal species to assure accurate measurements. Prior to every ICP-MS analysis, a performance report was executed to ensure that the instrument was working properly. All standard chemicals used in this study were trace metal grade. New calibration curves were created every six samples to correct any possible instrument drift in ICP-MS analysis. Also, the Collision Cell Technology (CCT) performance option was used to avoid the argon-based interferences in all ICP-MS analyses.

7.2.5 Statistical Analysis

Descriptive statistical analysis was applied to describe the range, mean, and standard deviation of the metal concentrations measured over the study period. Analysis of variance (ANOVA) was employed to quantify the difference in metal and nutrients concentrations measured between the sampling periods. A level of significance of 0.05 (α) was employed in the ANOVA analysis. Pearson correlation analysis was used to evaluate the correlations among the water quality parameters measured. All data were tested for normality and/or homogeneous variances using Minitab 15 (Minitab, Inc., State College, PA) before the statistical analyses.

7.3 Results

7.3.1 *Stream Flow and Rainfall*

The highest rainfall occurred in September (Fig. 7.2). The occurrence of the real peak stream flow was not known since the continuous daily stream flow data were not available. However, the limited stream flow data collected on the sampling dates basically followed the trend of the rainfall for the study period, i.e., an increase in rainfall induced an increase in the stream flow. Thus, the water and sediment samples collected on the day after a major rainfall event such as the sampling date in October may be influenced to some extent by surface runoff.

7.3.2 *Conventional Parameters*

The overall water depth of Manadas Creek was 1–1.5 ft during the study period. The average water temperatures ranged from 15.8 ± 0.0 °C to 31.1 ± 0.1 °C ($n = 10$). High temperatures occurred in July and August, and low temperatures occurred in January and February. The average water pH measured monthly ranged from 7.3 ± 0.01 to 8.0 ± 0.01 ($n = 10$). All pH values recorded met the U.S.EPA freshwater aquatic life criteria (6.5–9.0) (U.S.IBWC, 2003). The DO demonstrated significant (based on ANOVA results) temporal variation with extremely low DO levels (below 5.0 mg/L) in the months of March (3.52 ± 0.02 mg/L), April (2.49 ± 0.05 mg/L), July (2.93 ± 0.09 mg/L), August (0.84 ± 0.07 mg/L), and September (1.75 ± 0.12 mg/L) as shown in Fig. 7.3. Therefore, it did not meet the designated criteria for aquatic life use as specified in U.S.IBWC (2003) at the sampling site during these months. The TDS concentrations measured at Manadas Creek ranged from 1,500 to 5,500 mg/L with the highest value occurring in June. The salinity concentrations measured also showed a significant temporal variation (based on ANOVA results) and ranged from 1.38 ± 0.01 to 4.69 ± 0.01 ppt ($n = 10$) with the highest salinity observed in June.

7.3.3 *Nutrients*

All nutrients varied monthly and significantly based on ANOVA results (Fig. 7.3). The highest concentrations of all nutrient species occurred during the month of August, when the DO level was the lowest. The nutrient criteria suggested by

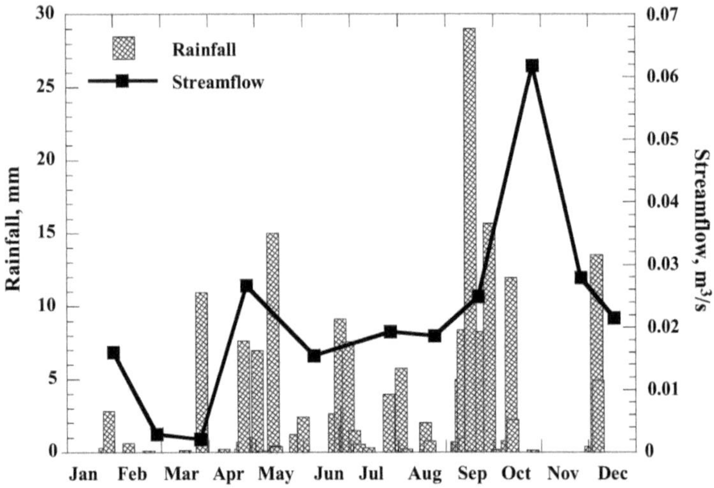

Fig. 7.2 Stream flow measured using a FlowTracker (SonTek/YSI, San Diego, CA, USA) on the sampling dates and daily rainfall data obtained from Texas Weather Connection website (http://twc.tamu.edu/)

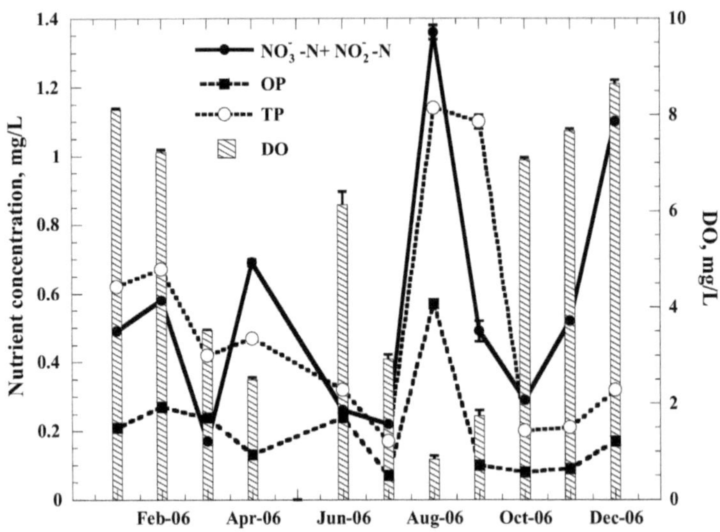

Fig. 7.3 Average nutrients and DO concentrations measured at Manadas Creek. *Error bars* represent ± one standard deviation of three measurements of each parameter at the two sub-locations

TCEQ before 2008 were 2.76 mg/L for the sum of NO_3^--N and NO_2^--N, 0.8 mg/L for TP, and 0.5 mg/L for OP in freshwater (TCEQ, 2004; U.S.IBWC, 2003). The new criteria listed in 2008 guidance are 1.95 mg/L for NO_3—N, 0.69 mg/L for TP, and 0.37 mg/L for OP in freshwater streams (TCEQ, 2008). Thus, Manadas Creek did not meet the criteria for TP in August and September and for OP in August in 2006.

7.3.4 *Dissolved Metals in the Stream Water*

Several dissolved trace metals including vanadium (V), Cr, Ni, copper (Cu), cobalt (Co), As, and Sb were found in Manadas Creek (Table 7.1). Sb, As, Cu, and Ni were found in higher concentrations compared to the other metals present at the sampling site. Sb was in high concentrations in July, October, and November and low concentration in December. For the months of August and September, As concentrations were high. The high concentrations of Cu were detected in August and September and for Ni in September and November. Despite these variations, ANOVA results only indicated a significant temporal variation for Ni and Co.

The recommended national water quality guide set by U.S.EPA specifies criteria for human health for the consumption of water and organisms, the Criteria Maximum Concentration (CMC) and the Criterion Continuous Concentration (CCC) for priority toxic pollutants (U.S.EPA, 2006). The CMC values represent the highest concentration of a material in surface water to which an aquatic community can be exposed briefly without resulting in an unacceptable effect. The CCC values represent the highest concentration of a material in surface water to which an aquatic community can be exposed indefinitely without resulting in an unacceptable effect. The available criteria for human health for the consumption of water and organisms listed by U.S.EPA are 5.6 μg/L for Sb, 0.018 μg/L for As and 1,300 μg/L for Cu. The available CMC values are 570 μg/L for Cr(III), 16 μg/L for Cr(VI), 470 μg/L for Ni, 340 μg/L for As, and 13 μg/L for Cu. The available CCC values are 74 μg/L for Cr(III), 11 μg/L for Cr(VI), 52 μg/L for Ni, 150 μg/L for As, and 9.0 μg/L for Cu. Vanadium (V) and Co values were not listed as a priority toxic pollutant by U.S.EPA. Thus, Manadas Creek exceeded the limits for human health for the consumption of water and organisms for Sb and As and the CMC and CCC values for Cu in the months of August and September.

7.3.5 *Sediment Associated Metals*

Several trace metals were found in the sediment in Manadas Creek (see Table 7.2). Concentrations of Sb were high compared to the other metals present in the sediment. Cr concentration was the next highest. ANOVA results indicated that no significant temporal variations exist for all metals detected in Manadas Creek sediment although the highest concentrations of Sb was found in October, which was 3.3 times the lowest concentrations detected in August and December. Cr concentrations were the highest for the month of October and lowest in July.

The criteria for sediment-associated metals for freshwater in streams is 7.0 mg/kg of dry sediment for As, 21.7 mg/kg of dry sediment for Cr, 14.5 mg/kg of dry sediment for Cu, and 15.5 mg/kg of dry sediment for Ni (TCEQ, 2004). The collected sediment samples met the TCEQ criteria for As, Cr, Cu, and Ni. Sb was not listed as a screening level metal in the sediment criteria guide suggested by TCEQ. However,

Table 7.1 Dissolved metal concentrations at Manadas Creek

Date M/D	V (μg/L)	Cr (μg/L)	Ni (μg/L)	Cu (μg/L)	Co (μg/L)	As (μg/L)	Sb (μg/L)
07/28	6.14 ± 0.70	1.66 ± 1.00	7.03 ± 0.71	3.67 ± 0.14	2.44 ± 0.23	7.53 ± 0.28	73.47 ± 5.70
08/25	3.49 ± 4.42	0.57 ± 0.81	1.50 ± 2.12	16.43 ± 5.15	0.44 ± 0.57	11.08 ± 15.50[d]	43.59 ± 48.36[d]
09/22	6.00 ± 0.52	6.13 ± 4.75	14.74 ± 0.05	11.85 ± 15.02	6.74 ± 1.13	27.72 ± 19.73[d]	37.91 ± 11.50
10/27	2.67 ± 0.04	2.79 ± 0.46	5.32 ± 0.82	4.48 ± 1.60	2.78 ± 0.29	3.49 ± 0.37	53.71 ± 0.55
11/27	4.66 ± 0.24	0.82 ± 0.15	10.78 ± 0.44	2.26 ± 0.51	2.36 ± 0.05	5.73 ± 0.95	56.36 ± 0.17
12/18	2.94 ± 1.20	0.72 ± 0.09	4.99 ± 2.00	5.40 ± 3.78	1.27 ± 0.49	2.22 ± 1.78	17.89 ± 11.12
Mean[a]	4.31	2.11	7.39	7.35	2.67	9.63	47.15
Range[b]	2.67–6.14	0.57–6.13	1.50–14.74	2.26–16.43	0.44–6.74	2.22–27.72	17.89–73.47
ANOVA results, F (F_{crit})[c]	1.28 (4.39)	2.24 (4.39)	**26.8 (4.39)**	1.37 (4.39)	**28.5 (4.39)**	10.4 (6.59)	1.44 (5.19)

Notes: Values reported represent mean ± one standard deviation of measurements from three replicates for each sample collected at the two sub-locations

[a]The arithmetic mean of the mean concentrations of each type of metal measured over time

[b]Range of the mean concentrations of each type of metal measured over time

[c]ANOVA analysis results among sampling periods. F_{crit} are critical values for $\alpha = 0.05$

[d]Data did not pass the normality ($p \geq 0.02$) and equal variance ($p \geq 0.04$) tests and were thus not included in ANOVA analysis

Table 7.2 Metals associated with sediments at Manadas Creek

Date M/D	V (mg/kg)	Cr (mg/kg)	Ni (mg/kg)	Cu (mg/kg)	Co (mg/kg)	As (mg/kg)	Sb (mg/kg)
07/28	0.06 ± 0.01	0.17 ± 0.04	0.02 ± 0.01	0.02 ± 0.01	0.01 ± 0.00	0.05 ± 0.02	1.55 ± 0.81
08/25	0.05 ± 0.04	0.21 ± 0.00	0.02 ± 0.01	0.01 ± 0.01	0.02 ± 0.00	0.05 ± 0.01	1.28 ± 0.77
09/22	0.05 ± 0.01	0.23 ± 0.02	0.03 ± 0.01	0.04 ± 0.03	0.03 ± 0.00	0.06 ± 0.01	2.92 ± 2.34
10/27	0.03 ± 0.01	0.64 ± 0.10	0.05 ± 0.04	0.04 ± 0.02	0.03 ± 0.02	0.03 ± 0.01	4.29 ± 1.01
11/27	0.03 ± 0.00	0.46 ± 0.18	0.07 ± 0.01	0.02 ± 0.00	0.01 ± 0.00	0.04 ± 0.01	3.53 ± 0.05
12/18	0.02 ± 0.01	0.32 ± 0.25	0.04 ± 0.01	0.05 ± 0.04	0.03 ± 0.03	0.06 ± 0.04	1.29 ± 0.68
Mean[a]	0.04	0.34	0.04	0.03	0.02	0.04	2.47
Range[b]	0.02–0.06	0.17–0.64	0.02–0.07	0.01–0.05	0.01–0.03	0.03–0.06	1.28–4.29
ANOVA results, F (F_{crit})[c]	1.89 (4.39)	3.73 (4.39)	2.20 (4.39)	0.97 (4.39)	0.81 (4.39)	0.80 (4.39)	2.43 (4.39)

Notes: Values reported represent mean ± one standard deviation of measurements from three replicates for each sample collected at the two sub-locations

[a]The arithmetic mean of the mean concentrations of each type of metal measured over time

[b]Range of the mean concentrations of each type of metal measured over time

[c]ANOVA analysis results among sampling periods. F_{crit} are critical values for $\alpha = 0.05$ and df = 11

Table 7.3 Correlation coefficients (R) among water quality parameters

	NO_3^--N + NO_2^--N	OP	TP	DO	TDS	pH	Q
Correlation matrix for nutrients in water							
NO_3^--N + NO_2^--N	1.000						
OP	***0.623***	1.000					
TP	0.516	***0.612***	1.000				
DO	−0.127	−0.294	***−0.538***	1.000			
TDS	−0.459	−0.301	−0.413	0.218	1.000		
pH	−0.226	−0.200	***−0.730***	***0.776***	0.173	1.000	
Q	−0.049	−0.408	−0.268	0.124	0.265	0.069	1.000

	V	Cr	Ni	Cu	Co	Q	TDS	pH
Correlation matrix for dissolved metals in water								
V	1.000							
Cr	0.468	1.000						
Ni	0.686	0.724	1.000					
Cu	−0.023	0.222	−0.197	1.000				
Co	0.601	***0.961***	***0.885***	0.048	1.000			
Q	−0.480	0.216	−0.050	−0.330	0.129	1.000		
TDS	−0.167	0.273	0.519	−0.550	0.384	0.654	1.000	
pH	−0.600	−0.595	−0.408	***−0.756***	−0.550	0.434	0.352	1.000

	V	Cr	Ni	Cu	Co	As	Sb	Q	TDS	pH
Correlation matrix for metals associated with sediments										
V	1.000									
Cr	−0.663	1.000								
Ni	−0.588	***0.750***	1.000							
Cu	−0.607	0.243	0.236	1.000						
Co	−0.443	0.080	−0.180	0.712	1.000					
As	0.056	***−0.750***	−0.473	0.374	0.432	1.000				
Sb	−0.269	***0.826***	***0.729***	0.177	0.008	***−0.730***	1.000			
Q	−0.414	***0.896***	0.446	0.253	0.207	***−0.758***	***0.821***	1.000		
TDS	−0.567	***0.817***	***0.917***	0.434	0.100	−0.476	***0.890***	0.654	1.000	
pH	−0.710	0.624	0.487	0.292	−0.128	−0.358	0.174	0.434	0.352	1.000

Note: Bold and italic correlation values are significant at $p < 0.1$ level; Q is stream flow

the background soil concentration data for Sb reported by the State of Texas is 1.1 mg/kg of dry sediment (U.S.EPA, 2005). Thus, this study indicates elevated levels of Sb in the Manadas Creek sediment, a cause for concern.

7.3.6 Correlations Among Water Quality Parameters

The sum of NO_3^--N and NO_2^--N was significantly correlated with OP, and TP was correlated with OP, DO, and pH in the creek water (Table 7.3). No significant correlations ($p < 0.1$) between stream flow and each of the nutrients in the stream were found. For dissolved metals, Cu was significantly correlated with pH, and Co was significantly correlated with Cr and Ni. Since data for As and Sb did not pass the normality test, they were not included in the correlation analysis. Similar to nutrients, no significant correlation ($p < 0.1$) was found between stream flow and dissolved metals. For sediment metals, the following groups of parameters were correlated at significant level ($p < 0.1$): Cr with Ni, As, Sb, stream flow, and TDS; Ni with Sb and TDS; As with Sb and stream flow; and Sb with stream flow and TDS (Table 7.3).

7.4 Discussion

The concentrations of Sb and As found in either the Manadas Creek sediments or water raise a concern according to the available criteria outlined by U.S.EPA (2005, 2006). Both As and Sb are naturally occurring elements found in mineral deposits. Anthropogenic sources of Sb include discharge from petroleum refineries, fire retardants, ceramics, and electronics and those of As include runoff from glass and electronics production wastes and fire retardants (U.S.EPA, 1997). Since Manadas Creek is situated in Laredo's industrial area, all of these potential anthropogenic sources could have contributed to the elevated levels of these two metals in the water and sediments samples.

The highest level of nutrients occurred in August; while, a significantly low DO of 0.84 ± 0.07 mg/L was observed during the same month. The close link between DO and nutrients levels can be illustrated via eutrophication, a process during which excessive plant growth and decay occur resulting in decreases in DO and severe reductions in water quality and fish and other animal populations (Bricker et al., 1999). During the sampling trip in August, decomposing algal blooms and small dead fish were present around the sampling site. Thus, the introduction of high levels of nutrients, algal blooms, and the August heat might result in low DO concentrations in the creek.

A previous study conducted by U.S.IBWC and TCEQ in June 1995 showed 0.13 mg/L NO_3^--N, 0.02 mg/L NO_2^--N, 0.16 mg/L TP, and 0.04 mg/L OP. All of these nutrients levels are lower than those detected in June of this study period, which were 0.21 ± 0.01 mg/L for NO_3^--N, 0.05 ± 0.00 mg/L for NO_2^--N, 0.32 ± 0.01 mg/L for TP, and 0.24 ± 0.00 mg/L for OP. Although these nutrients levels met the freshwater criteria suggested by TCEQ (TCEQ, 2004; U.S.IBWC, 2003) in June, they were not meeting the criteria for TP in August and September and for OP in August. This result indicates the importance of the time period selected for assessing water quality levels in Manadas Creek. In addition, it should be noted that a wastewater treatment plant is located 5.64 km upstream of the sampling site. The wastewater treatment plant effluent is used to irrigate the Country Club Golf Course (Fig. 7.1). Effluent that is not used by the golf course is released into Manadas Creek. During the dry season, this water becomes an important fraction of the total flow in Manadas Creek, a variable which might have contributed to the elevated nutrients levels observed in this study.

The comparison of metal levels detected in this study with those obtained before 2006 is summarized in Table 7.4. The concentrations for V and Co were not reported in the two phase binational studies conducted by U.S.IBWC and TCEQ in March 1993 and June 1995 (TNRCC, 1997; U.S.IBWC, 1994). For the water samples, the highest concentrations detected for Ni, Cr, Cu, As, and Sb in this study were higher than those detected in 1993 (U.S.IBWC, 1994) and 1995 (TNRCC, 1997). In the sediment samples the concentrations of all metals were found to be lower than those detected in 1993 and 1995. However, these differences should be examined carefully since sediments were extracted using a dilute HCl solution in the two phase binational studies, while a sequential extraction method of Silveira et al. (2006) was used in this study.

Only extractable or non-residual fraction of metals associated with sediments were analyzed in this study. Thus, the metal levels reported here only represent a small fraction of the total sediment metals occurring in Manadas Creek. The impact analyses conducted based on these results might have underestimated the potential negative effects of the sediment metals reported here since the sediment criteria used here and specified by U.S. EPA (2005) and TCEQ (2004) are total sediment metals. The sequential extraction technique is a frequently used approach to evaluate metal distribution in different chemical forms present in a solid phase (Silveira et al., 2006). It can provide useful information to predict the fate and the bioavailable fraction of heavy metals in the environment. However, determining total sediment metals would make it possible to gain a complete assessment of the metal contamination levels and should be a part of a future study.

Strong correlations were found between the measured stream flow data and sediment associated Sb, As, and Cr levels at Manadas Creek. However, no such significant correlations were found between the stream flow and the conventional parameters, dissolved metal concentrations in the stream water and nutrient levels. The correlation between trace metals and stream discharge has been evaluated by many researchers in many natural riverine systems, for example, Sherrell and Ross (1999). Particularly, the changes in hydrologic flow after rainfall events, which carries metal-rich waters and is flushed into streams, have been widely used to explain short-term increases in stream water metal concentrations in small watersheds (e.g., Cook et al., 1994). Recall that the sediment samples were collected from the first 6″ of the streambed in this study, representing the recently deposited fine sediments. Also, the stream flow data collected on the sampling dates at this site essentially followed the rainfall patterns as shown in Fig. 7.1, which implied the potential influence of the surface runoff resulting from rainfall events on the water quality parameters measured. Therefore, the strong positive correlations between the stream flow and sediments associated Sb and Cr indicate a net accumulation of these two metals in the streambed due to a potentially high replenishment rate resulting from surface runoff during rainfall events. In contrast, the strong negative correlation between the stream flow and the sediments associated As indicates a net loss of As from the streambed during rainfall events.

Table 7.4 Comparison of results with findings obtained before 2006

Metals	V	Cr	Ni	Cu	Co	As	Sb
U.S.IBWC (1994): *sampling time – March 1993; Location – 0.77 km downstream of the site sampled in this study*							
Water[a] (μg/L)	N/A[b]	<3.6	<4.7	1.6	N/A	<10	78
Sediment[c] (mg/kg)	N/A	11.1	10.6	8.46	N/A	5.2	N/A
TNRCC (1997): *sampling time – June 1995; Location – 0.77 km downstream of the site sampled in this study*							
Water (μg/L)	N/A	<2.0	<5.0	<7.0	N/A	6.3	36.4
Sediment (mg/kg)[c]	N/A	16.7	10.7	7.2	N/A	7.6	15.4
This study: *sampling time – July – Dec., 2006*							
Water (μg/L)	2.67–6.14	0.57–6.13	1.50–14.74	2.26–16.43	0.44–6.74	2.22–27.72	17.89–73.47
Sediment (mg/kg)[d]	0.02–0.06	0.17–0.64	0.02–0.07	0.01–0.05	0.01–0.03	0.03–0.06	1.28–4.29

[a]Dissolved phase metal
[b]Not available
[c]Extractable or non-residual fraction determined using diluted HCl
[d]Exchangeable sediments determined using sequential extraction method of Silveira et al. (2006)

The observed correlations among nutrients indicate that the occurrence of all nutrients may be due to a similar external source or they may have a similar fate and transport pattern in the creek water. Similarly, the significant correlations observed among dissolved and sediment metals indicate that Co, Cr, Ni, and Sb detected might have the same external source or a similar fate and transport pattern at the site studied here. Further investigation on the factors controlling the trace metals and nutrients distribution in Manadas Creek are needed in order to justify the above statements.

Dissolved As and Sb showed substantial temporal variations in the stream water (Table 7.1). Since they did not pass the normality/equal variance tests and were not included in the statistical analysis, ANOVA results only indicate a significant temporal variation for Ni and Co detected in the water phase. The temporal variation of metals in natural streams is governed by a wide range of stream processes such as watershed lithology, hydrology, pollution sources, and in-stream geochemical processes. Ni and Co were significantly correlated in the water column but they were not correlated with the stream flow (Table 7.3). Further research relating to other stream processes such as in-stream geochemical processes and watershed lithology in addition to the pollution sources will provide more insight on the causes of the observed significant temporal variations of these two metals. Moreover, additional data will be needed to further elucidate the potential temporal variations of As and Sb in the creek water.

Although results presented here help to fill in the scientific gaps of the water and sediment quality in the Laredo area, the data presented is quite limited. First, only one site was studied at Manadas Creek which limited our capability to evaluate the spatial distribution of the concerned water and sediment quality parameters and their interrelations. Additional spatial distribution of heavy metals and nutrients data, particularly along Manadas Creek and at its confluence with the Rio Grande, will provide valuable information on the impact of Manadas Creek water quality on the Rio Grande main channel. As a result, a follow up study focusing on spatial distribution of antimony and arsenic in Manadas Creek was conducted in 2008 (Baeza et al., 2010). The results of Baeza et al. provided solid evidence on the extent of the impact of past smelting activity on the creek water and sediment quality. Second, additional temporal data collected in the winter and during and right after the peak flow events will help address the complete seasonal variations and the effects of surface runoff and rainfall on the potential environmental problems of interest. Third, only dissolved phase metal concentrations were measured in the stream water in this study. The importance in distinguishing between dissolved and particulate metals to better interpret pollutant transport data collected in the field in surface water systems has been recognized by many researchers, e.g., Sherrell and Ross (1999), Yeats and Bewers (1982), and Taylor et al. (2012). To better assess the heavy metal contamination levels and gain more in-depth understanding of their fate and transport properties, total metal concentrations, which include both the dissolved phase and solid phase metal concentrations, should also be measured in future studies (e.g., Baeza et al., 2010). Finally, since the primary focus of this study was to provide preliminary assessment of the trace element levels in Manadas Creek, no major metal elements such as Fe, Al, and Si and sediment properties such as sediment sizes and organic matter content were measured. Future studies including this information will help to elucidate the mechanisms controlling the observed temporal variation of trace metals such as Co and Ni and the correlations among trace elements and nutrients.

7.5 Summary and Conclusions

This project was initiated to examine the water quality of Manadas Creek, an urban tributary of the Rio Grande in the Laredo, Texas area. Conventional water quality parameters (pH, DO, TDS, salinity, and temperature), sediment and water metal contents, and nutrients were determined from January to December 2006 at two sub-locations along the creek. Rainfall and stream flow data were also collected and presented.

The DO, TDS, and nutrients all showed significant temporal variations based on ANOVA results. The critical period of this project was the summer months of August and September when peak concentrations of nutrients and extremely low DO values were observed in the water. TP levels in August and September and OP in August exceeded the TCEQ's freshwater criteria. The low DO values observed in March, April, July, August, and September were below the designated use criteria set by U.S.IBWC for aquatic life.

Several trace metals including V, Cr, Ni, Cu, Co, As, and Sb were found in the Manadas Creek's water and sediments samples. Sb and As had elevated concentrations in the aqueous phase according to the U.S.EPA national recommended water quality criteria (U.S.EPA, 2006). Cu concentrations slightly exceeded the CMC and CCC values set by U.S.EPA in the months of August and September. The concentrations of Sb found in the Manadas Creek sediments raise a concern according to the reported Texas background Sb soil concentration data (U.S. EPA, 2005). In addition, strong correlations were found between the measured stream flow data and the sediments associated As, Sb, and Cr levels, among the metals detected, and among the nutrients detected.

Acknowledgements This material is based upon work supported by the Center of Research Excellence in Science and Technology – Research on Environmental Sustainability of Semi-Arid Coastal Areas (CREST-RESSACA) at Texas A&M University-Kingsville (TAMUK) through a Cooperative Agreement (No. HRD-0206259) from the National Science Foundation (NSF). Any opinions, findings, and conclusions or recommendations expressed in this material are those of the authors and do not necessarily reflect the views of the National Science Foundation. We would like to thank Elizabeth Cortez, an undergraduate research assistant, for helping with setting-up equipment and collecting data. We would also like to thank Andrew G. Smith for reviewing the manuscript before submission.

References

Agency for Toxic Substances, Disease Registry [ATSDR]. (2005). *Toxicological profile for nickel.* Washington, DC: US Public Health Service.

American Public Health Association [APHA]. (1992). *Standard methods-for the examination of water and wastewater* (18th ed.). Washington, DC: American Water Works Association/Water Environment Federation.

Baeza, M., Ren, J., Krishnamurthy, S., & Vaughan, T. (2010). Spatial distribution of antimony and arsenic levels in Manadas Creek, an urban tributary of the Rio Grande in Laredo, Texas. *Archives of Environmental Contamination and Toxicology, 58*(2), 299–314. doi:10.1007/s00244-009-9357-0.

Bricker, S., Clement, C., Pirhalla, D., Orlando, S., & Farrow, D. (1999). *National estuarine eutrophication assessment: A summary of conditions, historical trends, and future outlook.* Silver Spring, MD: Special Projects Office in cooperation with National Centers for Coastal and Ocean Science, National Ocean Service, and National Oceanic and Atmospheric Administration.

Cook, P., Jolly, I., Leaney, F., & Walker, G. (1994). Unsaturated zone tritium and chlorine 36 profiles from southern Australia: Their use as tracers of soil water movement. *Water Resource Research, 30*, 1709–1719.

Flores, B., Camarena, C., Ren, J., Krishnamurthy, S., & Belzer, W. (2009). Assessment of organochlorine pesticides levels in Manadas Creek, an urban tributary of the Rio Grande in Laredo, Texas. *Archives of Environmental Contamination and Toxicology, 57*(1), 11. doi:10.1007/s00244-008-9237-z.

HACH. (2003). *DR/2500 spectrophotometer procedure manual.* Loveland, CO: HACH Company.

Mendoza, J., Botsford, J., Hernandez, J., Montoya, A., Saenz, R., Valles, A., et al. (2004). Microbial contamination and chemical toxicity of the Rio Grande. *BMC Microbiology, 4*(17), 1–16.

Sherrell, R., & Ross, J. (1999). Temporal variability of trace metals in New Jersey pinelands streams: Relationships to discharge and pH. *Geochimica et Cosmochimica Acta, 63*(19–20), 3321–3336.

Silveira, M., Alleoni, L., O'Connor, G., & Chang, A. (2006). Heavy metal sequential extraction methods – A modification for tropical soils. *Chemosphere, 64*, 1929–1938.

Taylor, H. E., Antweiler, R. C., Roth, D. A., Alpers, C. N., & Dileanis, P. (2012). Selected trace elements in the Sacramento River, California: Occurrence and distribution. *Archives of Environmental Contamination and Toxicology, 62*(4), 557–569.

Texas Commission on Environmental Quality [TCEQ]. (2003). *Surface water quality monitoring procedures: Vol. 1 Physical and chemical monitoring methods for water, sediment and tissue.* Austin, TX: Texas Commission on Environmental Quality.

Texas Commission on Environmental Quality [TCEQ]. (2004). *Guidance for assessing Texas surface and finished drinking water quality data.* Austin, TX: TCEQ Office of Compliance and Enforcement, Monitoring Operations Division, Surface Water Quality Monitoring Program.

Texas Commission on Environmental Quality [TCEQ]. (2008). *Guidance for assessing and reporting surface water quality in Texas.* Austin, TX: TCEQ Office of Compliance and Enforcement, Monitoring Operations Division, Surface Water Quality Monitoring Program.

Texas Natural Resource Conservation Commission [TNRCC]. (1997). *Second phase of the binational study regarding the presence of toxic substances in the Rio Grande/Rio Bravo and its tributaries along the boundary portion between the United States and Mexico* (Final report). Austin, TX: Texas Natural Resource Conservation Commission.

TNRCC. (2002). *State of the Rio Grande and the environment of the border region, strategic plan: Vol. 3 Fiscal years 2003–2007.* Austin, TX: Texas Natural Resource Conservation Commission.

United States Environmental Protection Agency [U.S. EPA]. (1997). *Damage cases and environmental releases from mines and mineral processing sites.* Washington, DC: Offices of Solid Waste.

United States Environmental Protection Agency [U.S. EPA]. (2005). *Guidance for developing ecological soil screening levels*. Washington, DC: Office of Solid Waste and Emergency Response.

United States Environmental Protection [U.S. EPA]. (2006). *National recommended water quality criteria: U.S. environmental protection agency, office of water*. Washington, DC: U.S. EPA.

United States Section, International Boundary and Water Commission [U.S.IBWC]. (1994). *Binational study regarding the presence of toxic substances in the Rio Grande/Rio Bravo and its tributaries along the boundary portion between the United States and Mexico*. El Paso, TX: U.S.IBWC.

United States Section, International Boundary and Water Commission [U.S.IBWC]. (2003). *Regional assessment of water quality in the Rio Grande basin*. El Paso, TX: U.S. International Boundary and Water Commission.

Yeats, P., & Bewers, J. (1982). Discharge of metals from the St. Lawrence River. *Canadian Journal of Earth Science, 19*, 982–992.

Chapter 8
Antimony and Arsenic Isotope Distribution in a Semi-Arid Creek During the Wet Season

Marcia Baeza, Jianhong Ren, Sushma Krishnamurthy, and Thomas C. Vaughan

Abstract Previous studies found elevated metal concentrations in Manadas Creek. In this study, samples were collected from six sites along the creek during the wet period of 2008 and analyzed for dissolved and total metal isotopes in water and total metal isotopes in sediments. Results were compared with those obtained in a dry period. It was found that ^{75}As concentration in water was typically the lowest following a major rain event. A similar trend was observed for antimony (Sb) at the two sites, close to a decommissioned Sb smelter. Mann-Whitney test indicates significant differences in the medians of ^{75}As concentrations observed in the sediment between the dry and the wet seasons at most sites and for ^{121}Sb at two sites sampled. The enrichment factors (EFs) calculated for As in sediment were low (close to 1); however, they were up to 1,530.9 and 1,519.1 for ^{121}Sb and ^{123}Sb, respectively. Cluster analysis grouped ^{75}As with ^{208}Pb, ^{118}Sn, ^{66}Zn, and ^{137}Ba, or with ^{55}Mn, or with ^{60}Ni depending on the sites sampled; however, both Sb isotopes were not grouped with any other metals. Pearson correlation only indicated significant and strong correlation between ^{75}As and sediment properties. Significant correlation between ^{75}As and other metals was only found at one site. Most of the correlations between ^{121}Sb/^{123}Sb and other metals were poor. These results contribute to the understanding of the seasonal variations of Sb and As contamination in a semi-arid creek.

Keywords Rio Grande • Manadas Creek • Water • Sediment • Metals

---

M. Baeza • J. Ren (✉)
Department of Environmental Engineering, Frank H. Dotterweich College of Engineering, Texas A&M University-Kingsville, Kingsville, TX 78363-8202, USA
e-mail: jianhong.ren@tamuk.edu

S. Krishnamurthy
School of Sciences, University of Louisiana at Monroe, Monroe, LA 71209, USA

T.C. Vaughan
Department of Biology and Chemistry, Texas A&M International University, Laredo, TX 78041, USA

D. Ramirez et al. (eds.), *Environmental Sustainability Issues in the South Texas–Mexico Border Region*, DOI 10.1007/978-94-007-7122-2_8,

8.1 Introduction

The Rio Grande, a natural boundary between the United States and Mexico, supports extensive agriculture, recreation and wildlife (Texas Natural Resource Conservation Commission [TNRCC], 1997). The study site, Manadas Creek (MC), is an urban tributary of the lower Rio Grande located in a semi-arid region of Laredo, Texas, where most of the rainfall occurs in June through October (Rappole et al., 1986). Warehouses, a cement ready-mix factory, a major railroad, and a defunct antimony smelter are located in the vicinity of the creek within the Manadas Creek watershed. The smelter was established in 1928 and used a blast furnace to process antimony-bearing ore to produce antimony oxide until 1992 (United States Environmental Protection Agency [U.S.EPA], 1997). The plant was decommissioned in 1999 but continues to house two slag sites with antimony byproducts for remediation (TNRCC, 2002). To prevent contaminants from entering the creek, a retention pond was constructed between the two slag sites. The construction of a bridge over the creek beginning in April 2006 resulted in extensive changes to the Manadas Creek study site adjacent to the smelter. Most of the vegetation along the banks of the creek was removed and replaced by concrete.

Manadas Creek was identified as a tributary with high potential for toxic chemicals such as heavy metals in a two-phase, binational study conducted by the International Boundary and Water Commission, United States Section (U.S.IBWC) in 1992–1993 and 1995 (International Boundary and Water Commission, United States Section [U.S.IBWC], 1994; TNRCC, 1997). Specifically, the concentration of antimony (Sb) in water was found to be as high as 78 μg/L, which exceeded the human health screening level by 5.6 times and the national 85th percentile by 1.4 times (U.S.IBWC, 1994). In 2006, the highest dissolved Sb concentration in water was found to be 78 μg/L in August, and sediment exchangeable Sb ranged from 0.73 to 5.01 mg/kg of dry weight (Flores, 2007). The highest dissolved arsenic (As) concentration was 42 μg/L (Flores, 2007). Due to the differences in sampling locations, sampling time, and analytical methods used in the previous studies, it was not possible to make meaningful comparisons and draw conclusions on the spatial and temporal variation of Sb and As in Manadas creek.

In 2008, water and sediment core samples were collected from six sites along the creek in February, April, and May and analyzed for dissolved and total metals in water and total metals in sediments by Baeza et al. (2010). Results from this study showed a clear spatial distribution of As and Sb for both water and sediment samples collected during a dry period of time. In particular there were high concentrations of metals in the sediment samples: 420 mg/kg of ^{121}Sb and 14.2 mg/kg of ^{75}As, which was found at the maximum sampled sediment depth of 35.0 cm. In the creek water, the dissolved phase As ranged from 1.7 to 20.6 μg/L with the highest concentration occurring at MC-1 in April. Sb ranged from 1.2 to 220.6 μg/L with the highest concentration occurring at MC-3 in April. This study offered valuable information on the As and Sb contamination levels and their spatial distributions in Manadas Creek with a focus on the dry season. However, information on the distribution of metals during the wet season is still lacking.

Semi-arid regions usually encounter severe erosion problems caused by wind and surface runoff (Chopin et al., 2003; Querol et al., 2000; Razo et al., 2004; Van Geen et al., 1999; Wulf et al., 2012). In this type of environment, heavy metals contained in the residues from mining and metallurgical operations are often dispersed by wind and/or water after their disposal (Adriano, 2001; Iavazzo et al., 2012; Kim et al., 2012). Thus, soil and mine spoil texture, landscape topography, and regional and local climate all have a significant effect on contaminant transport in semi-arid regions (Bertine et al., 1978; Kim et al., 2012; Navarro et al., 2008). Distribution of metals in semi-arid creeks is usually related to physical processes (e.g. physical mixing in the downstream direction, geomorphic processes, and recent fluvial history) (Graf et al., 1991; Kim et al., 2012). Since these physical processes may change rapidly over time, their effect on the transport of metals in semi-arid regions implies a potentially significant seasonal variation of metal contamination.

Semi-arid creeks are characterized by seasonal rainfall and intermittent stream flow. The contaminant load in semi-arid creeks often occurs as "flushing and dilution" caused by these intermittent rain events (Ben-Othman et al., 1997; Ellis, 1986). In general, storms of longer duration generate more dilute runoff (Brezonik & Stadelmann, 2002). Rapid contaminant buildup usually occurs during dry periods since stream flow and dilution caused by rain events are limited (Brezonik & Stadelmann, 2002; Vaze & Chiew, 2002). These highly dynamic metal behaviors may further complicate the assessment of the impact of heavy metal contamination in these creeks. This complexity also makes it difficult to apply information on fate, transport, and distribution of heavy metals obtained from one creek to other creeks.

The objective of this study was to evaluate the seasonal variation of Sb and As concentrations in Manadas Creek and to contribute to the understanding of metal behaviors in semi-arid creeks. To achieve this objective, samples of water and sediment were obtained during the wet months in August, September, and November of 2008 and the concentrations of metals were compared with those obtained in the dry months (Baeza et al., 2010) for the same sampling sites. The results presented will provide insight on the underlying processes affecting temporal Sb and As distribution in a semi-arid creek.

8.2 Methods and Materials

8.2.1 Description of the Sampling Sites

Six sites were selected along the creek with two located upstream (MC-1 and MC-2) and four downstream (MC-3 to MC-6) of the antimony smelter. These sites were the same ones as sampled by Baeza et al. (2010). Table 8.1 shows the geographic location for each site, the distance measured from site MC-1, the maximum sediment depth sampled, and the water depth measured during each sampling trip. A detailed map for the study site and sampling locations can be found in (Baeza et al., 2010).

MC-1 was used as a control site since metal concentrations at this site were lower than those at the sites located downstream of the smelter (Baeza et al., 2010). MC-2 is located 2.61 km downstream of MC-1 and next to a cement ready-mix plant adjacent to the frontage road on the east side of Interstate Highway 35 (I-35). MC-3 and MC-4 are the sites closest to the decommissioned antimony smelter. Construction of the Mines Road overpass near these two sites took place between April 2006 and October 2008. MC-5 changed considerably during the wet season. On August 15th, 2008, the 8 m wide access road began washing away, leaving large slabs of asphalt at the bottom of the creek. By November 15th, the access road was completely washed away, making accessing to the site a difficult task. MC-6 is located 0.77 km downstream of the antimony smelter and 0.8 km upstream of the confluence with the Rio Grande. The creek bed at this location is composed of limestone; thus, only water samples were collected from this site. The width and surroundings of each site were the same as those reported by Baeza et al. (2010). The water level at each site had a minimal increase during the wet season compared to those found in the dry season.

8.2.2 Sample Collection

Water and sediment samples were collected on August 15, September 27, and November 15, 2008. Additional water samples were collected on August 18, 2008 after a 66.6 mm rain event on August 17, 2008. Precipitation records and weather data were obtained from the National Oceanic and Atmospheric Administration (NOAA) website (www.srh.noaa.gov) for Station of Laredo, TX (LRD) (27° 31′ N and 99° 28′ W). Water quality parameters measured in-situ included conductivity, pH, salinity, temperature, and turbidity. Turbidity was measured using a Turbidimeter Hach 2100P (Hach Company, Loveland, CO) and the others were measured using a YSI 63 (YSI Inc., Yellow Springs, OH). Stream velocity was recorded using an Acoustic Doppler Velocimeter (ADV) FlowTracker (SonTek/YSI, San Diego, CA) at a single point close to the channel center of each sampling site.

Water and sediment samples were collected following procedure recommended by the Texas Commission on Environmental Quality (TCEQ) (Texas Commission on Environmental Quality [TCEQ], 2003). Water samples were collected 6 in. below the water surface at midstream using a 1-L polyethylene dipper and stored in 1-L amber plastic bottles. Two sediment cores were extracted at each site using a gravity core sampler (Aquatic Research Instruments, Hope, ID) and kept in an upright position to preserve the vertical sediment profiles during transportation. The maximum depth of core penetration (Table 8.1) depended on the amount of

Table 8.1 Description of the sampling sites along Manadas Creek

Sites	Latitude (N)	Longitude (W)	Distance from site MC-1 (km)	Maximum depth of sediment core (cm)	Water depth (m)		
					August 15, 2008	September 27, 2008	November 15, 2008
MC-1 (Control)	27° 35′ 44.16″	−99° 28′ 41.16″	0	10	0.45	0.45	0.77
MC-2	27° 34′ 58.08″	−99° 29′ 56.04″	2.61	15	0.54	0.60	0.68
MC-3	27° 34′ 32.16″	−99° 30′ 21.24″	4.07	45	1.17	1.18	0.94
MC-4	27° 34′ 29.28″	−99° 30′ 22.68″	4.17	40	0.95	0.95	0.90
MC-5	27° 34′ 30.36″	−99° 30′ 34.20″	4.54	25	1.25	0.92	0.87
MC-6	27° 34′ 28.92″	−99° 30′ 45.36″	4.84	No sediment was collected	0.35	0.45	0.40

sediment available at each site. All water and sediment samples were placed on ice after collection and immediately transported to the laboratory for further treatment.

8.2.3 Sample Analysis

On arrival at the laboratory, sediment cores were maintained in an upright position and preserved at 4 °C. The following day sediment cores were sliced every 5 cm using a sediment extruder (Aquatic Research Instruments, Hope, ID). Moisture and organic carbon content were determined on each 5-cm sample interval based on a loss on ignition method using a split of each sample interval after sample homogenization (Dean & Dean, 1974). The remaining sediment of each 5-cm sample interval was then homogenized in 250-mL precleaned wide-mouth amber glass jars, air dried at ambient conditions in plastic trays for 2–3 days, crushed, and air-dried again (Baeza et al., 2010). For sediment metal analysis, a fraction of 0.5 g of the prepared dry sediment was digested following EPA method 3051A (U.S.EPA, 2007), preserved with 2 % HNO_3 and analyzed using an Inductively Coupled Plasma-Mass Spectrometry (ICP-MS) (Thermo Electron, Xseries, UK) following method 200.8 (U.S.EPA, 1994) at Texas A&M University-Kingsville (TAMUK).

Water samples were prepared for dissolved metals analysis by filtering samples using 0.45 μm filters and digested for total metals analysis using method 3015A (U.S.EPA, 2007). All processed samples were preserved with 2 % HNO_3, and analyzed using an ICP-MS (Thermo Electron, Xseries, UK) following method 200.8 (U.S.EPA, 1994) at TAMUK. All metal results were reported as metal isotopes with atomic number indicated as a superscript in front of each metal element.

8.2.4 Data Analysis

Statistical analyses (cluster analysis and Pearson correlation analysis) for metals detected were performed in the same fashion as explained by Baeza et al. (2010). Cluster analysis was performed for sediment samples collected in November using individual metal enrichment factor (EF) obtained at each depth. The EF values were calculated according to the method of Brady (1984) using MC-1 as the control site and Fe as a normalizing element via the equation $EF = (M/Fe)_{sample}\left[(M/Fe)_{ref}\right]^{-1}$, where M_{sample} and M_{ref} are the concentrations of targeted metals in the sediment samples collected at each study site. Pearson correlation analysis was conducted using the individual sediment metal concentration and sediment properties (moisture content and organic matter content) collected at each depth in November. All data were tested for normality and/or homogeneous variances using Minitab 15 (Minitab, Inc., State College, PA) before the statistical analyses. Only normal data sets were used in the cluster and correlation analysis presented here. In addition, the Mann-Whitney test, a non-parametric significance test, was performed to evaluate the difference of metal concentrations obtained in the dry and wet periods.

8.2.5 Quality Control Measures

Sample cross contamination was avoided by using new pairs of latex gloves at each collection site. Plastic containers were pre-cleaned with 10 % hydrochloric acid and rinsed in deionized (DI) water before use. For each sample preparation in the laboratory, new disposable plastic pipette tips were used. Triplicate analyses were conducted for sediment moisture content and organic carbon content, and duplicate analyses were carried out for water and sediment sample metal analyses. Indium was used as an internal standard in all ICP-MS analyses. The percentage recovery for all analytes of interest was in a range of 80–115 %. All metals reported here were within the detection limits of the ICP-MS. All blanks, calibration standards, and fortified samples were prepared according to the EPA 200.8 method. These quality control measures were also employed by Baeza et al. (2010).

8.3 Results

8.3.1 Conventional Water Quality Parameters

January through June is considered the dry season since rainfall is typically minimal during this period (Fig. 8.1). However, a 43.43 mm rain event was recorded on May 5, 2008. The rain events are typically more frequent from July through December. The most rain occurred in August 2008, with a monthly total of 180.8 mm. The heaviest rainfall event with a total of 66.6 mm rain occurred from August 17 through August 18, right before the sample collection on August 18. The water velocity measured on August 15, September 27, and November 15 was proportional to the rain events and was the highest in September due to the occurrence of rain events prior to the sampling trip. Even though the highest rain event occurred on August 17, stream velocity was not recorded on the August 18 sampling trip since the ADV FlowTracker was not available during this sampling trip. Thus, the velocity measured on August 15 represented the stream velocity condition prior to the major rainfall events occurring in August. Stream velocity during September ranged between 0.2 and 7.8 cm/s at sites MC-1 through MC-5, and had a high value of 40.9 cm/s at MC-6, which was primarily due to the decreased cross sectional area at this site. In November, a higher stream velocity was measured at site MC-5 than at MC-6, which might be due to the slightly increased channel slope and shallower water depth observed at site MC-5 during this sampling trip.

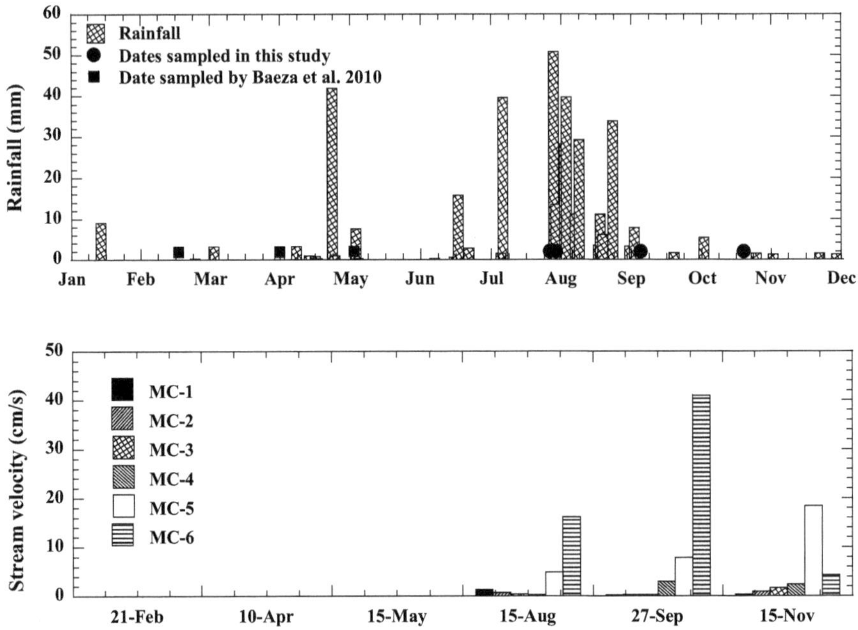

Fig. 8.1 Rainfall and stream velocity at Manadas Creek. The dates sampled for the dry season by Baeza et al. (2010) were February 21, April 10, and May 15. No velocity data were collected on February 21, April 10 and August 18 since the ADV flowtracker was not available

The water temperature was in the range of 27.2–29.3 °C in August, 22.1–23.4 °C in September, and 17.4–18.5 °C in November 2008. A pH range of 6.6–8.0 was recorded during the wet season and at all sites. The lowest pH at all sites was reported on August 18th, 2008 at MC-6. Conductivity values were higher during the dry months (January through June) than those recorded during the wet months (July through December) (Fig. 8.2). The lowest conductivity values were reported on August 18, which is attributed to the dilution of salts and other chemical components due to the rain events recorded prior to the sampling (Fig. 8.1). The conductivity increased in September and November. Turbidity was only measured in the wet season. The maximum turbidity of 113.00 ± 4.80 NTU was recorded at MC-3 on August 15. Turbidity along the creek was fairly constant around 40 NTU on August 18 and decreased at most sampling sites in September and November.

8.3.2 Sediment Properties

During the wet months, the depth-averaged moisture content ranged from 0.23 ± 0.00 to 0.97 ± 0.08, the porosity ranged from 0.38 ± 0.00 to 0.69 ± 0.10, and the organic matter content (OM%) ranged from 1.02 ± 0.29 to 6.64 ± 0.26 % at all sites. Note that the moisture content, w, is defined as the ratio

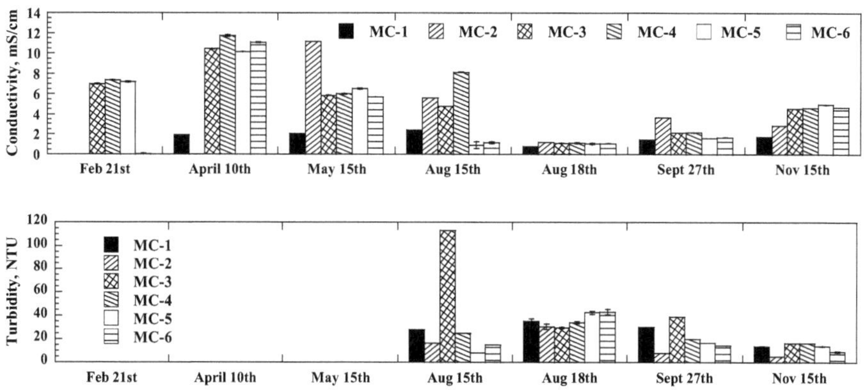

Fig. 8.2 Conductivity and turbidity at sites MC-1 through MC-6. *Error bars* represent ± one standard deviation (SD) of the measurements conducted. Conductivity data for the dry season (January through June) are from (Baeza et al., 2010)

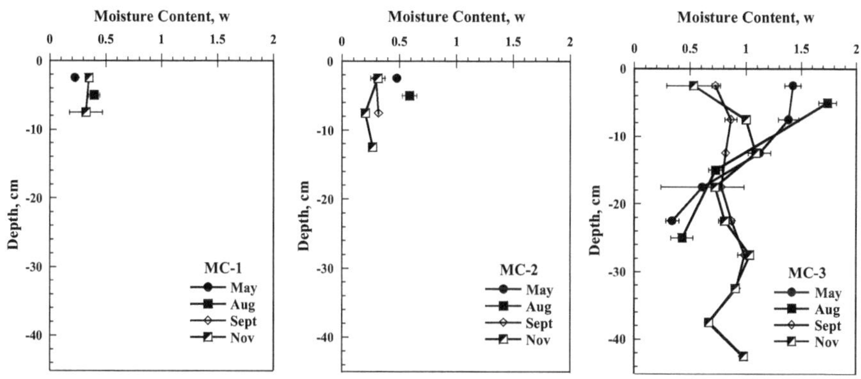

Fig. 8.3 Moisture content as a function of depth for sediments collected at sites MC-1 through MC-3. Values reported for May are from Baeza et al. (2010). *Error bars* represent ± one standard deviation (SD) of the measurements conducted for different sediment cores

of weight of water to the weight of dry sediment. The vertical profiles of the moisture content observed at MC-1, MC-2, and MC-4 were similar in range at all sampling depths and times (Figs. 8.3 and 8.4). The surface sediments in the cores collected during May (Baeza et al., 2010) and August at MC-3 had a high moisture content, which decreased with increasing depth. Moisture content measured at MC-3 in September and November was fairly constant with the depth. Moisture content at MC-5 was fairly constant with depth for August and May samples. However, for September and November samples, it increased slightly with depth and the maximum value was reported to be 0.69 ± 0.02 at a depth of 17.5 cm in the core collected in November.

MC-1 and MC-2 reflected fairly constant concentrations of OM% at all depths and time with higher OM% observed in August and September and the lowest OM%

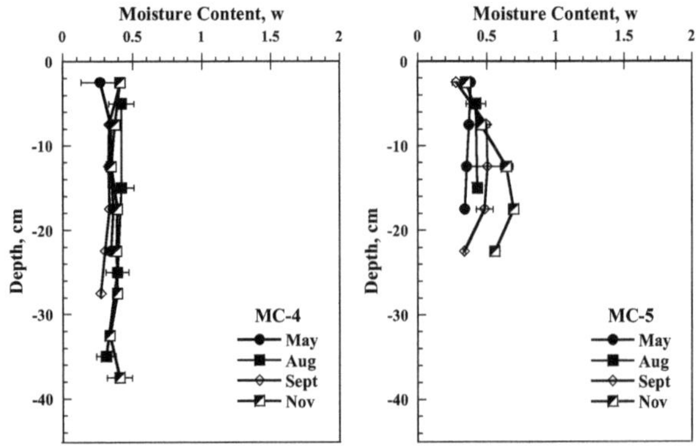

Fig. 8.4 Moisture content as a function of depth for sediments collected at sites MC-4 and MC-5. Values reported for May are from Baeza et al. (2010). *Error bars* represent ± one standard deviation (SD) of the measurements conducted for different sediment cores

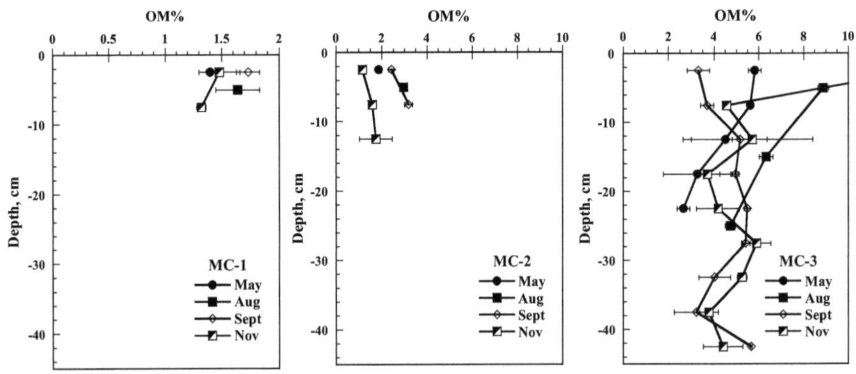

Fig. 8.5 OM% as a function of depth for sediments collected at sites MC-1 through MC-3. Values reported for May are from Baeza et al. (2010). *Error bars* represent ± one standard deviation (SD) of the measurements conducted for different sediment cores

occurring in November (Figs. 8.5 and 8.6). OM% at MC-3 had the highest values among all sites. It was fairly constant in September and was high in surficial sediments and decreased with depth in May, August and November. OM% had an almost constant vertical distribution at sites MC-4 and MC-5. The lowest OM% at MC-4 was found in November and highest in August. At MC-5, OM% in November increased with depth. Black layers of sediment were visible when the sediment cores collected at this site were extruded. These black layers might be caused by the decomposition of vegetation and other entrapped materials due to the washing away of a paved road following large rain events.

8.3.3 Metal Concentrations in the Creek Water

Water samples were analyzed for ^{75}As, ^{121}Sb, and ^{123}Sb for both dissolved and total phases. Particulate phase metals were calculated from the difference between the

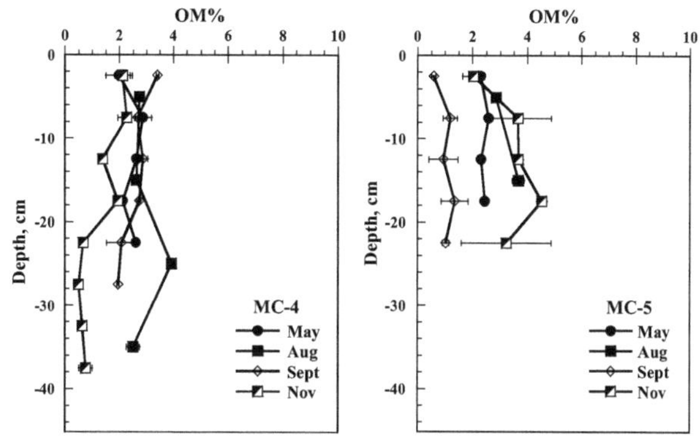

Fig. 8.6 OM% as a function of depth for sediments collected at sites MC-4 and MC-5. Values reported for May are from Baeza et al. (2010). *Error bars* represent ± one standard deviation (SD) of the measurements conducted for different sediment cores

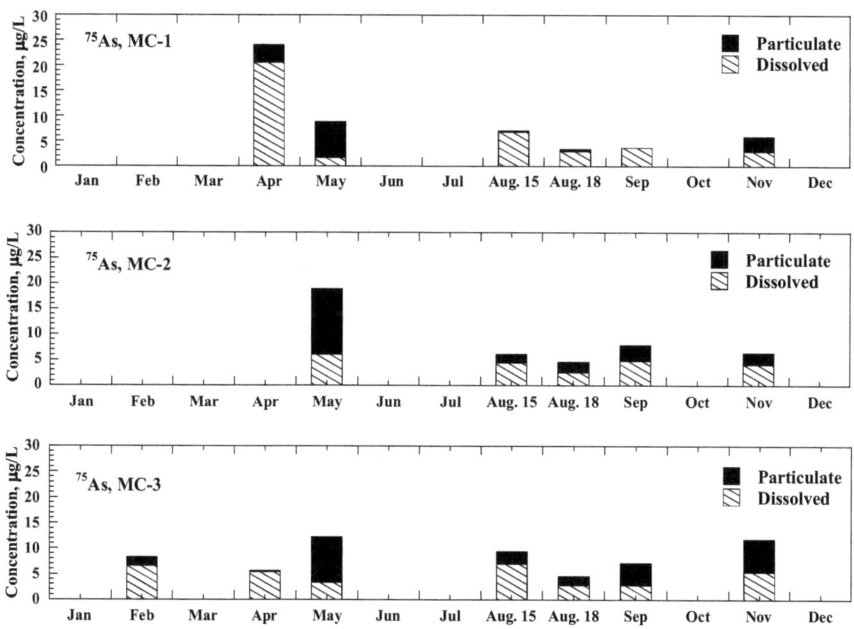

Fig. 8.7 As concentration in water samples collected during 2008 at sites MC-1 through MC-3. Results for February, April, and May are from Baeza et al. (2010). Note MC-1 and MC-2 were not sampled in February, 2008

total and dissolved phase metals. Results of the non-parametric significance test, Mann-Whitney test, showed that there were no significant ($\alpha > 0.05$) differences between the medians of ^{75}As concentrations observed in the water phase in the dry season and those obtained in the wet season at all sites sampled. However, Figs. 8.7 and 8.8 show that As levels peaked in either April or May at all sites and decreased to its lowest levels on August 18th after the rain event on August 17th. The concentrations increased again in September and November. Particulate As concentration was the highest in May at MC-1 through MC-4 and in November at MC-5 and MC-6.

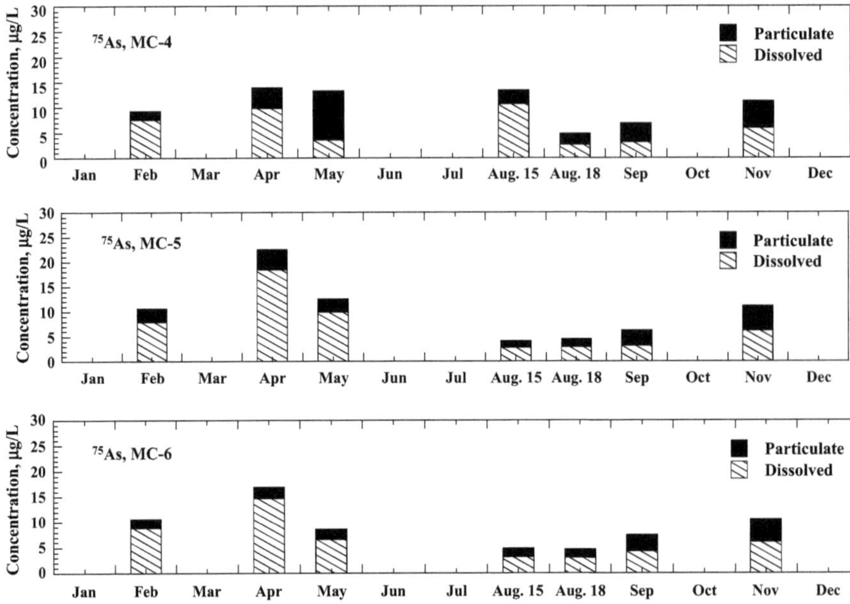

Fig. 8.8 Water phase As concentration distribution during 2008 at sites MC-4 through MC-6. Results for February, April, and May are from Baeza et al. (2010)

Since the two Sb isotopes (^{121}Sb and ^{123}Sb) had similar concentrations (the mass ratio of ^{121}Sb/^{123}Sb was around 1) among all sampling sites and dates; thus, only the results of ^{121}Sb are presented here. As shown in Figs. 8.9 and 8.10, ^{121}Sb concentrations at sites MC-1 and MC-2 increased slightly after the rain event on August 17th, and then decreased in September and November. The highest ^{121}Sb concentration was reported on August 15th for MC-3, in April for MC-4 and MC-5, and in February at MC-6. ^{121}Sb concentrations were typically higher during the dry months compared to the wet months. They decreased after the August 17 rain event at sites MC-3 (tenfold) and MC-4; however, ^{121}Sb concentration increased at sites MC-5 and MC-6 after the rain event. At sites MC-3 through MC-6, ^{121}Sb concentration gradually increased in the samples collected during September and November, following the rain event. However, Mann-Whitney test indicates that there were no significant ($\alpha > 0.05$) differences between the medians of ^{121}Sb concentrations observed in the dry season and those obtained in the wet season at all sites sampled.

8.3.4 Vertical Distribution of Metals in the Sediments

The vertical distribution of ^{75}As was fairly constant at all depths for all sites, and Mann-Whitney test indicates that there were significant ($\alpha < 0.05$) differences between the medians of ^{75}As concentrations observed in the dry season and those

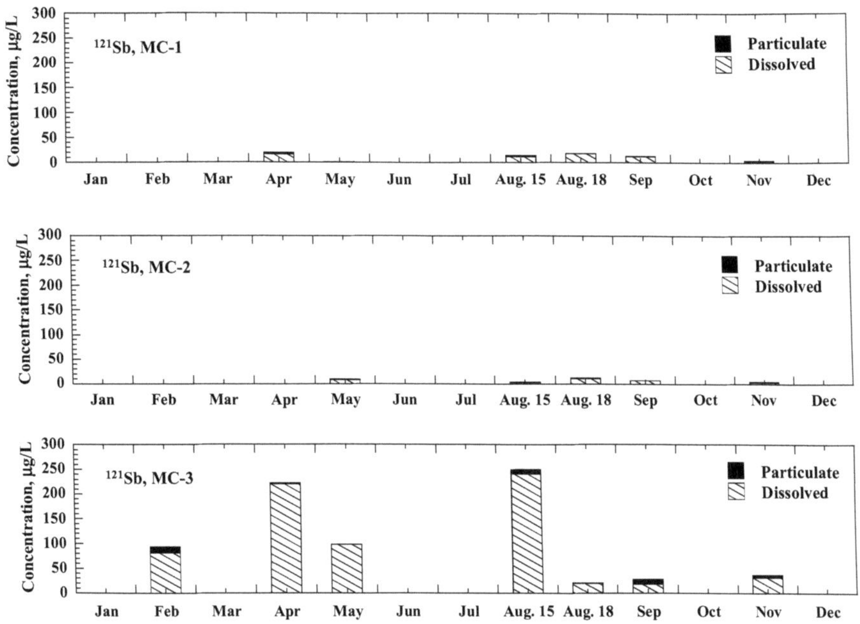

Fig. 8.9 ^{121}Sb concentration in water samples collected during 2008 at sites MC-1 through MC-3. Results for February, April, and May are from Baeza et al. (2010). Note MC-1 and MC-2 were not sampled in February, 2008

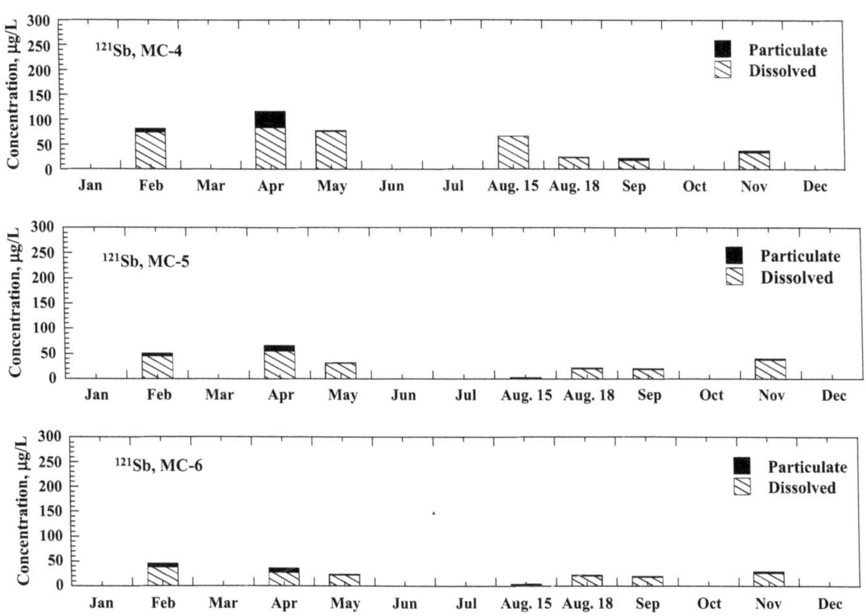

Fig. 8.10 Water phase ^{121}Sb concentration distribution during 2008 at sites MC-4 through MC-6. Results for February, April, and May are from Baeza et al. (2010)

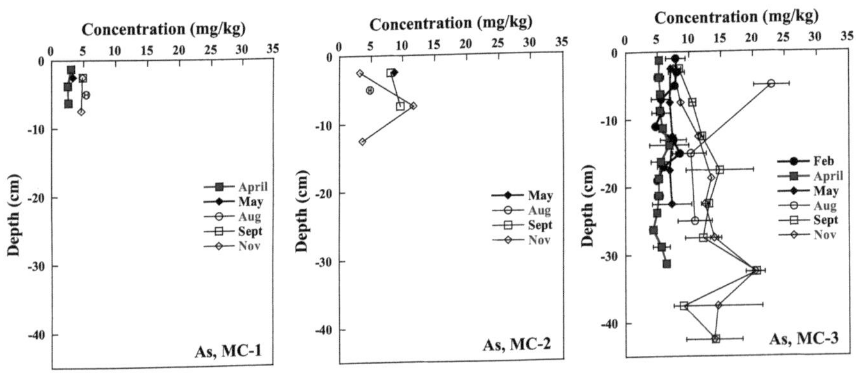

Fig. 8.11 Distributions of ^{75}As concentration in the creek sediments at sites MC-1 through MC-3. Data shown are the average metal concentrations measured from all sediment cores sampled at each site. Results for February, April, and May are from Baeza et al. (2010). *Error bars* represent ± 1 SD of measurements obtained from all sediment cores sampled at each site

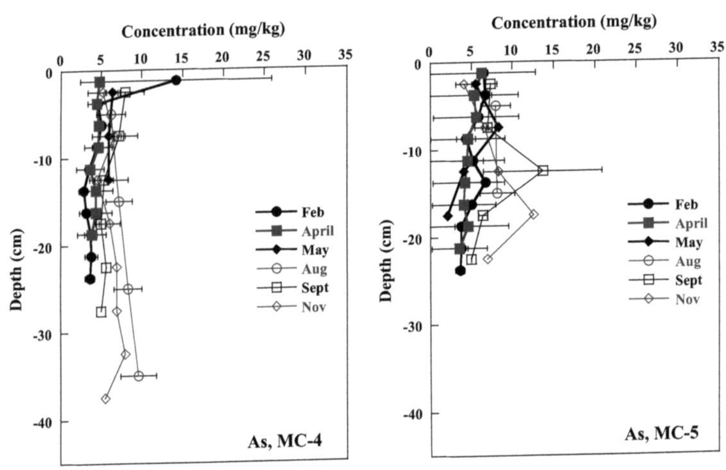

Fig. 8.12 Distributions of ^{75}As concentration in the creek sediments at sites MC-4 and MC-5. Data shown are the average metal concentrations measured from all sediment cores sampled at each site. Results for February, April, and May are from Baeza et al. (2010). *Error bars* represent ± 1 SD of measurements obtained from all sediment cores sampled at each site

obtained in the wet season at all sites sampled except site MC-2 (Figs. 8.11 and 8.12). The As concentration detected at MC-3 showed higher concentrations for the wet season at all depths. The highest concentration was found in September (20.39 mg/kg) and November (20.56 mg/kg) at a depth of 32.5 cm. The background soil concentration data for As reported by the State of Texas is 6.4 mg/kg of dry sediment (U.S.EPA, 2005). Consequently, the As concentration exceeded the Texas background soil As concentration at MC-3 for all depths in August, September, and November; MC-4 at 15–35 cm in August, 0–12.5 cm in September, and all depths in November; and MC-5 at depths 0–15 cm in August and September and 7.5–22.5 cm

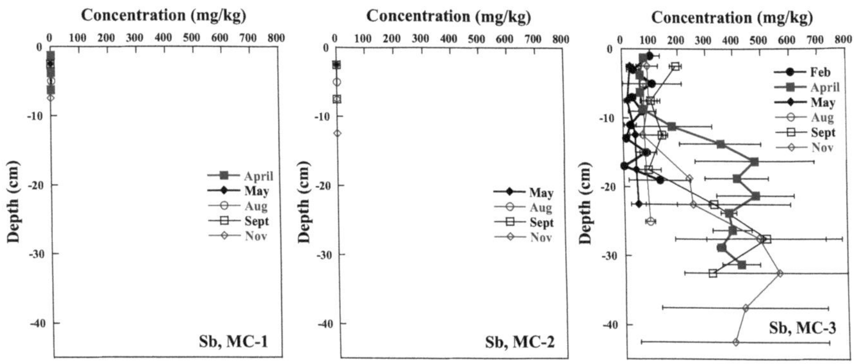

Fig. 8.13 Distributions of ^{121}Sb in the creek sediments at sites MC-1 through MC-3. Data shown are the average metal concentrations measured from all sediment cores sampled at each site. Results for February, April, and May are from Baeza et al. (2010). *Error bars* represent ± 1 SD of measurements obtained from all sediment cores sampled at each site

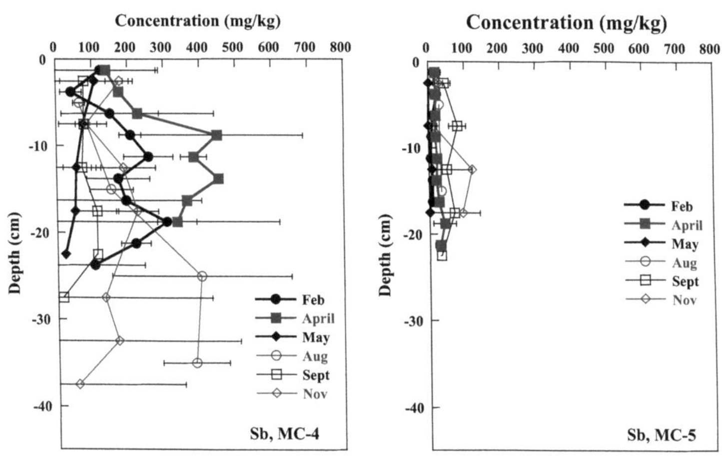

Fig. 8.14 Distributions of ^{121}Sb in the creek sediments at sites MC-4 and MC-5. Data shown are the average metal concentrations measured from all sediment cores sampled at each site. Results for February, April, and May are from Baeza et al. (2010). Error bars represent ± 1 SD of measurements obtained from all sediment cores sampled at each site

in November. The findings at MC-4 and MC-5 are comparable with what was reported for the dry season (Baeza et al., 2010). However, the Texas background soils As concentrations at site MC-4 were not exceeded during the dry months.

Low concentrations of ^{121}Sb in both dry and wet seasons were observed at MC-1 and MC-2, and Mann-Whitney test results indicate no significant variations ($\alpha > 0.05$) were found between the sampling months (Figs. 8.13 and 8.14). These ^{121}Sb concentrations did not exceed the Texas background Sb soil concentration data of 1.1 mg/kg of dry sediment (U.S.EPA, 2005). At MC-3, ^{121}Sb had a higher concentration in April, September, and November than was reported in the

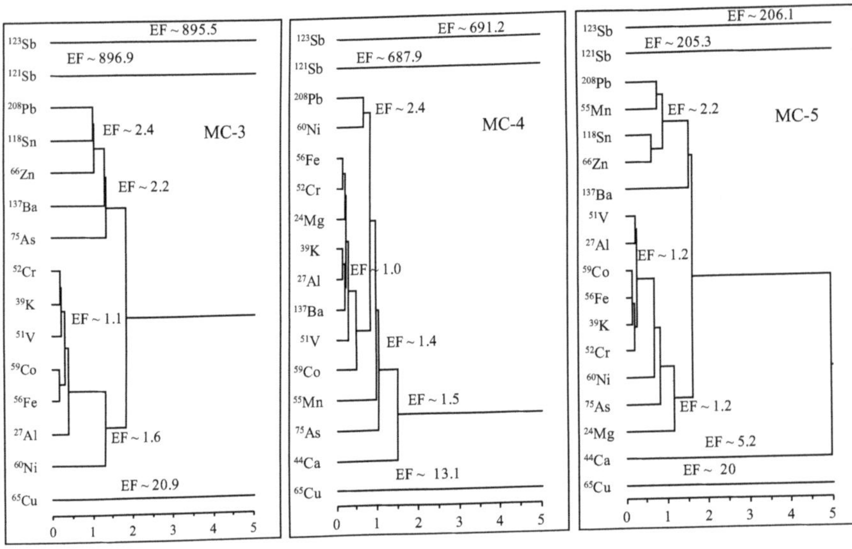

Fig. 8.15 Dendrograms of metal EF values obtained in sediments collected in November at sites MC-3 through MC-5. EF values labeled are the depth-averaged EF values; however, individual EF values calculated at each depth were used in the cluster analysis. Note: x-axis represents the distance or similarity between clusters

remaining sampling months. A maximum concentration of 552.12 mg/kg occurred in November at a depth of 32.5 cm. This level of ^{121}Sb is about 500 times higher than the Texas background Sb soil concentration level. MC-4 had a higher ^{121}Sb concentration in April and August than was reported in other sampling months. At MC-5, Sb concentrations in August were lower compared to those recorded for September and November. In addition, much lower Sb concentrations with insignificant vertical variations were observed on all sampling dates at MC-5 compared to the other sites. For the wet months, all ^{121}Sb concentrations observed at sites MC-3 through MC-5 were higher than the reported Texas background Sb soil concentration level. Mann-Whitney test indicated that there were significant ($\alpha < 0.05$) differences between the medians of ^{121}Sb concentrations observed in the dry season and those obtained in the wet season at sites MC-3 and MC-5.

8.3.5 Cluster and Correlation Analysis for Metals Detected in the Creek Sediments

Figure 8.15 shows the dendrograms of metal Enrichment Factors (EFs) obtained in sediments collected in November at sites MC-3, MC-4, and MC-5. The dendrograms show the grouping of metals with similar EF values. For sites MC-3 through MC-5, the depth-averaged EF values calculated for ^{75}As ranged from 1.2 to 2.2, for ^{121}Sb ranged from 205.3 to 896.9, and for ^{123}Sb ranged from 206.1 to 895.5

with the highest EF values occurring at MC-3. The EF range reported for ^{75}As is similar to the range (1.0–1.7) for samples collected in May, 2008 (Baeza et al., 2010); while the maximum EF value reported for ^{121}Sb was about 12 times larger at MC-3, 4 times larger at MC-4, and 25 times larger at MC-5 than those were reported by Baeza et al. (2010) for May, 2008 (EF = 74.5 at MC-3, 205.1 at MC-4, and 8.2 at MC-5).

At MC-3, metals including ^{52}Cr, ^{39}K, ^{51}V, ^{59}Co, ^{56}Fe and ^{27}Al, with an EF value close to 1.1 were grouped together in November sediments; while in May sediments, ^{84}Sr, ^{84}Kr, ^{52}Cr, ^{60}Ni, ^{51}V, ^{27}Al, ^{118}Sn, ^{58}Ni, and ^{56}Fe, were grouped together at this site with EF values near 1 (Baeza et al., 2010). EFs calculated for ^{75}As at MC-3 in November ranged from 1.5 to 2.7 at different depths and they ranged from 203.1–1530.9 to 205.3–1519.1 for ^{121}Sb and ^{123}Sb, respectively. At site MC-3 in the wet season, ^{75}As was grouped with ^{208}Pb, ^{118}Sn, ^{66}Zn, and ^{137}Ba; while in the dry season, ^{75}As was not included in the cluster analysis at MC-3. Similar to the results in May, both Sb isotopes were not grouped with any other metals in the wet season.

At MC-4, ^{56}Fe, ^{52}Cr, ^{24}Mg, ^{39}K, ^{27}Al, ^{137}Ba ^{51}V, and ^{59}Co were grouped together with an EF value approximate to 1. ^{75}As was grouped with ^{55}Mn with an EF value of 1.4 in the sediment core collected in November, while, it was grouped with Cu and Pb in May sediments. Similar to results obtained at MC-3 in this study and by Baeza et al. (2010) at MC-4, both Sb isotopes were not grouped with any other metals. A similar grouping pattern was observed at MC-5 except that the depth-averaged EF values for Sb isotopes (205.3 for ^{121}Sb and 206.1 for ^{123}Sb, respectively) were much lower at site MC-5 than those reported at other sites, a similar pattern that was reported by Baeza et al. (2010) for the dry months.

The correlation between ^{75}As and sediment properties was only significant ($p \leq 0.05$) and strong (R = 0.91 for w and R = 0.92 for OM%) at MC-5 (Table 8.2), which is stronger than those reported in May sediments (R = 0.74 for w and R = 0.28 for OM%, $p > 0.05$). Weak and insignificant correlations (R < 0.71 and $p > 0.05$) were found between the Sb isotopes (^{121}Sb and ^{123}Sb) and sediment properties (w and OM%). The correlation between ^{75}As and ^{24}Mn, ^{118}Sn, ^{137}Ba, ^{121}Sb, ^{123}Sb, and ^{208}Pb was significant at MC-3 (R in the range of 0.67–0.87, $p \leq 0.05$). No significant correlation ($p \leq 0.05$) was found between ^{75}As and any other metals at MC-4 and MC-5. Most of the correlations between ^{121}Sb/^{123}Sb and other metals were poor (R ≤ 0.80). These findings were similar to those reported by Baeza et al. (2010).

8.4 Discussion and Conclusions

The level of metal contamination was determined to understand its seasonal variation in a semi-arid creek. Results were compared with those reported for the dry period (Baeza et al., 2010). It was found that the highest ^{75}As concentration in water occurred in either April or May (the dry months) at all sites. It decreased on August

Table 8.2 Correlation matrix (R) for sediment metals and properties measured in November

MC-3				MC-4				MC-5			
Metal	^{75}As	^{121}Sb	^{123}Sb	Metal	^{75}As	^{121}Sb	^{123}Sb	Metal	^{75}As	^{121}Sb	^{123}Sb
^{24}Mg	0.62	0.55	0.55	^{24}Mg	0.10	0.33	0.33	^{24}Mg	0.71	0.47	0.47
^{27}Al	0.56	0.54	0.54	^{27}Al	0.00	0.52	0.52	^{27}Al	0.71	0.48	0.48
^{39}K	0.63	0.63	0.63	^{39}K	0.03	0.48	0.48	^{39}K	0.70	0.47	0.47
^{44}Ca	0.57	0.50	0.50	^{44}Ca	0.06	−0.01	0.00	^{44}Ca	0.57	0.34	0.35
^{51}V	0.47	0.42	0.42	^{51}V	0.13	0.57	0.56	^{51}V	0.70	0.44	0.44
^{52}Cr	0.64	***0.69****	***0.70****	^{52}Cr	−0.08	0.31	0.31	^{52}Cr	0.80	0.58	0.58
^{55}Mn	***0.73****	***0.71****	***0.71****	^{55}Mn	0.01	−0.15	−0.15	^{55}Mn	0.67	0.65	0.65
^{56}Fe	0.62	0.55	0.56	^{56}Fe	0.12	0.21	0.20	^{56}Fe	0.72	0.36	0.36
^{59}Co	0.61	0.57	0.58	^{59}Co	−0.27	−0.16	−0.16	^{59}Co	0.78	0.42	0.42
^{60}Ni	0.49	0.63	0.63	^{60}Ni	0.26	0.06	0.06	^{60}Ni	0.60	0.22	0.23
^{65}Cu	0.60	0.62	0.63	^{65}Cu	−0.17	−0.26	−0.26	^{65}Cu	0.63	0.22	0.22
^{66}Zn	0.57	0.62	0.62	^{66}Zn	0.22	0.45	0.44	^{66}Zn	0.57	0.49	0.49
^{75}As	1.00	***0.87****	***0.87****	^{75}As	1.00	−0.25	−0.25	^{75}As	1.00	0.64	0.64
^{118}Sn	***0.69****	***0.77****	***0.78****	^{118}Sn	−0.19	0.69	0.69	^{118}Sn	0.86	0.78	0.78
^{121}Sb	***0.87****	1.00	***1.00****	^{121}Sb	−0.25	1.00	***1.00****	^{121}Sb	0.64	1.00	***1.00****
^{123}Sb	***0.87****	***1.00****	1.00	^{123}Sb	−0.25	***1.00****	1.00	^{123}Sb	0.64	***1.00****	1.00
^{137}Ba	***0.67****	***0.79****	***0.79****	^{137}Ba	0.13	0.19	0.19	^{137}Ba	0.60	0.10	0.10
^{208}Pb	***0.70****	***0.80****	***0.80****	^{208}Pb	0.01	***0.70****	***0.70****	^{208}Pb	0.73	0.83	0.83
w	0.21	0.11	0.11	w	−0.30	−0.24	−0.24	w	***0.91****	0.71	0.71
OM%	−0.50	−0.37	−0.37	OM%	−0.42	0.40	0.40	OM%	***0.92****	0.52	0.52

*Indicates significant correlation ($p \leq 0.05$)

18th after a major rain event and increased gradually in September and November. A similar change was observed for ^{121}Sb at MC-3 and MC-4; however, ^{121}Sb concentration slightly increased at MC-1 and MC-2 after a major rain event in August and decreased in September and November. At MC-5 and MC-6, ^{121}Sb concentration increased after a major rain event and in September and November.

The decrease in ^{75}As (all sites) and ^{121}Sb (at MC-3 and MC-4) and increase in ^{121}Sb (at MC-5 and MC-6) concentrations after rain events might indicate that metals were flushed away from the creek, transported downstream, and/or diluted due to the increased stream flow. The increase of ^{121}Sb concentration at the upstream sites (MC-1 and MC-2) after the rain events might be due to the introduction of ^{121}Sb by surface runoff or atmospheric deposition because release from the bed sediment should be inconsequential because the sediment ^{121}Sb concentration at these sites was low. Due to its possible enhancement in road strength, the antimony smelter's by products were used to cover roads in the surrounding area as a local practice, which makes the leaching of ^{121}Sb at sites MC-1 and MC-2 via surface runoff during rainfall possible. The increase of ^{75}As and ^{121}Sb in September and November at sites MC-3 through MC-6 suggests that ^{75}As and ^{121}Sb were gradually released from the bed sediments at these sites.

Mann-Whitney test indicated significant ($\alpha < 0.05$) differences in the medians of ^{75}As concentrations observed in the sediment between the dry and the wet seasons at all sites except MC-2 and for ^{121}Sb at sites MC-3 and MC-5. ^{121}Sb was at its highest in November at MC-3 with a concentration of 552.12 mg/kg of dry sediment. The calculated EF values indicated that ^{121}Sb and ^{123}Sb occurring at sites MC-3 through MC-5 was substantially greater than the background concentration, and it might have been introduced by an anthropogenic source as also indicated by Baeza et al. (2010). The main difference between the dry and wet months was that the EF values reported for ^{121}Sb and ^{123}Sb at MC-3 through MC-5 were lower in May than those reported in November. The EF values for ^{75}As were low in both dry and wet months, indicating a lesser extent of ^{75}As contamination at the sites downstream of the defunct antimony smelter as compared to ^{123}Sb and ^{121}Sb.

Sediment metal vertical profiles reflected heterogeneity at sites MC-3 and MC-4 as also concluded by Baeza et al. (2010). There was an increase in sediment ^{121}Sb concentration in April, September, and November at MC-3 and in April and August at MC-4, but there is not enough evidence to attribute this increase to rain events or physical processes. Therefore, further research is needed to determine the controlling factors of ^{121}Sb mobilization in the sediment. In both dry and wet seasons, sediment Sb concentrations at MC-3 through MC-5 exceeded the Texas background Sb concentration of 1.1 mg/kg of dry sediment at all depths.

To date no studies with a focus on seasonal variations of metals in Manadas Creek have been conducted. The binational study was carried out during dry seasons at one site (half-way between MC-5 and MC-6) on the creek (TNRCC, 1997; U.S.IBWC, 1994). Flores (2007) studied metal distribution from July to December 2006, but was limited to only one site (half-way between MC-3 and MC-4). Figure 8.16 compares our findings with data gathered in these previous studies. During the wet season for this study, the dissolved As concentration in water was lower than those reported in the two-phase, binational study and also by Baeza et al. (2010) for MC-5 and MC-6 (U.S.IBWC, 1994; TNRCC, 1997). It was also slightly lower than those reported by Baeza et al. (2010) and Flores (2007) for sites MC-3 and MC-4 (Fig. 8.16a). As concentrations in sediments obtained in this study were similar to those reported in TNRCC (1997), higher than that reported in U.S.IBWC (1994) and by Baeza et al. (2010) at MC-5. They are much higher than those reported by Baeza et al. (2010) and Flores (2007) at sites MC-3 and MC-4 (Fig. 8.16b).

Dissolved Sb concentrations in water obtained in this study were lower than that reported by U.S.IBWC (1994), TNRCC (1997), and Baeza et al. (2010) for sites MC-5 and MC-6. At sites MC-3 and MC-4, they were within a similar range to those reported by Flores (2007) and much lower than those reported by Baeza et al. (2010). The Sb concentrations observed in the sediment samples in this study were higher than those reported by Baeza et al. (2010), U.S.IBWC (1994), and TNRCC (1997) at site MC-5. At sites MC-3 and MC-4, they were similar to those reported by Baeza et al. (2010) and much higher than those reported by Flores (2007). It should be noted that the treatment methods used to determine metal concentration in the sediments were different for the aforementioned studies. U.S.IBWC (1994) and TNRCC (1997) reported extractable or non-residual fraction of As and Sb concentrations in the sediment determined using diluted HCl. Flores (2007) reported exchangeable sediments using the sequential extraction method of Silveira et al. (2006). Results presented in this paper and by Baeza et al. (2010) correspond to

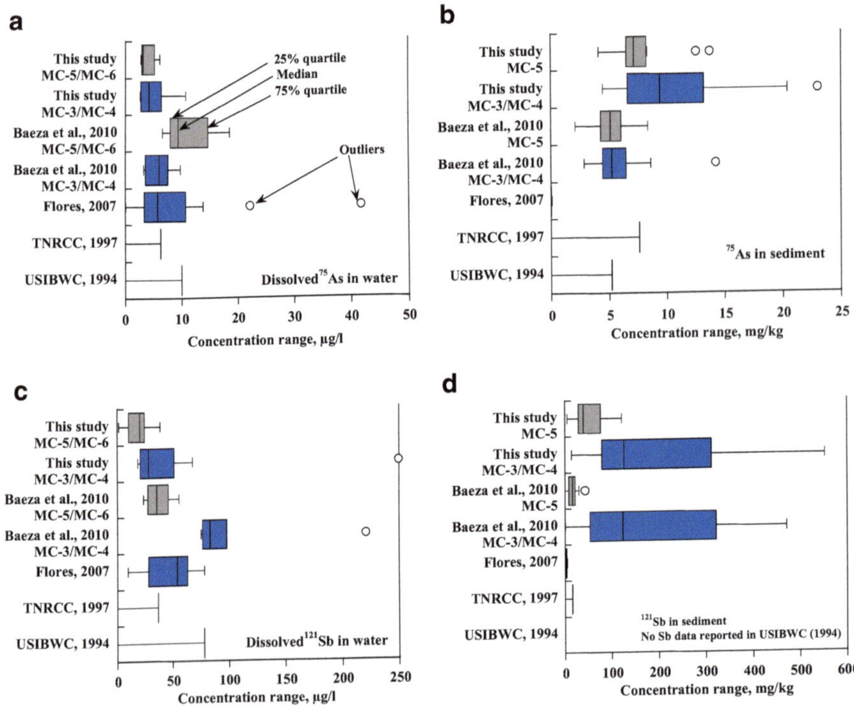

Fig. 8.16 Comparison of results with previous findings. (**a**) Dissolved As concentrations in water, (**b**) As concentrations in sediment, (**c**) dissolved Sb concentrations in water, and (**d**) Sb concentrations in sediment. U.S.IBWC (1994) and TNRCC (1997) studies were conducted at a site half-way between MC-5 and MC-6. Flores (2007) study was conducted at a site half-way between MC-3 and MC-4

total metal concentration in sediments via microwave digestion. Therefore, due to the differences in methods the comparison between the sediment results reported here and those obtained by previous authors should be carefully evaluated.

Statistical analysis for the wet season confirmed the results on As and Sb grouping and their correlations with sediment properties and other metals presented in Baeza et al. (2010). Sb and As in the sediment did not decrease substantially during the wet season as was hypothesized. Therefore, the metals deposited in the sediment were either only slightly mobilized during the wet season or not mobilized at all. As stated by Graf et al. (1991) for streams in semi-arid environments, the distribution of metals in the streambed might be affected mainly by physical processes (e.g., surface runoff, rainfall events, and flow rate). Therefore, the intensity of the rain might have a great effect on the mobilization of metals stored in the streambed. This assumption has yet to be proven. Further research should focus on the impact of rain intensity on sediment mobilization at Manadas Creek since many of the sites often present low water levels and mixing or resuspension of surficial sediments is possible.

The results presented here show that As contamination in the sediment may not exhibit a major problem at Manadas Creek due to its low EF values (close to 1).

However, Sb concentration in the sediment is extremely high compared to the Texas background soil concentration, which makes the continuous release of Sb to the stream water possible. This potential mobilization of Sb could lead to contamination problems in an extended area downstream of the decommissioned antimony smelter. It could also be transported into the Rio Grande mainstream and cause intermittent impacts on the water quality of the Rio Grande. Short-term and long-term impacts caused by physical processes (e.g., surface runoff, rainfall events, and flow rate) on the water quality of Manadas Creek and the Rio Grande should be assessed in future studies for a better understanding of the controlling mechanisms of Sb mobilization and transport in this creek and other similar semi-arid creeks.

Acknowledgements This article was based upon work supported by the Center of Research Excellence in Science and Technology–Research on Environmental Sustainability of Semi-Arid Coastal Areas (CREST–RESSACA) at Texas A&M University-Kingsville (TAMUK) through a Cooperative Agreement (HRD-0734850) from the National Science Foundation (NSF). It was also supported by a TAMUK Graduate Assistantship and Scholarship from The National Council on Science and Technology (CONACYT) of Mexico to M. Baeza. Any opinions, findings, and conclusions or recommendations expressed in this material are those of the authors and do not necessarily reflect the views of the National Science Foundation or the National Council on Science and Technology of Mexico. We would like to thank Andrew G. Smith for reviewing the manuscript before submission.

References

Adriano, D. (2001). *Trace elements in terrestrial environments: Biogeochemistry, bioavailability and risks of metals* (2nd ed.). New York: Springer.

Baeza, M., Ren, J., Krishnamurthy, S., & Vaughan, T. (2010). Spatial distribution of antimony and arsenic levels in Manadas Creek, an urban tributary of the Rio Grande in Laredo, Texas. *Archives of Environmental Contamination and Toxicology, 58*(2), 299–314. doi:10.1007/s00244-009-9357-0.

Ben-Othman, D., Luck, J., & Tournoud, M. (1997). Geochemistry and water dynamics: Application to short time scale flood phenomena in a small Mediterranean catchment. *Chemical Geology, 140*, 9–28.

Bertine, K., Walawender, S., & Koide, M. (1978). Chronological strategies and metal fluxes in semi-arid lake sediments. *Geochimica et Cosmochimica Acta, 42*, 1559–1571.

Brady, N. (1984). *The nature and properties of soils*. New York: MacMillan.

Brezonik, L., & Stadelmann, H. (2002). Analysis and predictive models of storm water runoff volume, loads, and pollutant concentrations from watersheds in the Twin Cities metropolitan area, Minnesota, USA. *Water Research, 36*, 1743–1757.

Chopin, E., Black, S., Hodson, M., Coleman, M., & Alloway, B. (2003). A preliminary investigation into mining and smelting impacts on trace element concentrations in the soils and vegetation around Tharsis, SW Spain. *Mineralogical Magazine, 67*, 279–288.

Dean, W., & Dean, W. (1974). Determinations of carbonate and organic matter in calcareous sediments and sedimentary rocks by loss on ignition: Comparison with other methods. *Journal of Sedimentary Petrology, 44*(1), 242–248.

Ellis, J. (1986). Pollution aspects of urban runoff. In H. C. Trono, J. Marsalek, & M. Desbordes (Eds.), *Urban runoff pollution NATO ADI series* (pp. 1–38). New York: Springer.

Flores, B. (2007). *Chemical contamination of the lower Rio Grande near Laredo, TX*. Unpublished master's thesis, Texas A&M University-Kingsville, Kingsville, TX.
Graf, W., Clark, S., Kammerer, M., Lehman, T., Randall, K., & Schroeder, S. (1991). Geomorphology of heavy metals in the sediments of Queen Creek, Arizona, USA. *Catena, 18*(6), 567–582.
Iavazzo, P., Ducci, D., Adamo, P., et al. (2012). Impact of past mining activity on the quality of water and soil in the high Moulouya Valley (Morocco). *Water, Air, and Soil Pollution, 223*(2), 573–589. doi:10.1007/s11270-011-0883-9.
Kim, C. S., Stack, D. H., & Rytuba, J. J. (2012). Fluvial transport and surface enrichment of arsenic in semi-arid mining regions: Examples from the Mojave Desert, California. *Journal of Environmental Monitoring, 14*(7), 1798–1813. doi:10.1039/c2em30135k.
Navarro, M., Perez-Sirvent, C., Martinez-Sanchez, M., Vidal, J., Tovar, P., & Bech, J. (2008). Abandoned mine sites as a source of contamination by heavy metals: A case study in a semi-arid zone. *Journal of Geochemical Exploration, 96*, 183–193.
Querol, X., Alastuey, A., López-Soler, A., & Plana, F. (2000). Levels and chemistry of atmospheric particulates induced by a spill of heavy metal mining wastes in the Doñana area, Southwest Spain. *Atmospheric Environment, 34*, 239–253.
Rappole, J. H., Russell, C. E., Norwine, J. R., & Fulbright, T. E. (1986). Anthropogenic pressures and impacts on marginal, neotropical, semiarid ecosystems: The case of South Texas. *Science of the Total Environment, 55*, 91–99.
Razo, I., Carrizales, L., Castro, J., Díaz Barriga, F., & Monroy, M. (2004). Arsenic and heavy metal pollution of soil, water and sediments in a semi-arid climate mining area in Mexico. *Water, Air, and Soil Pollution, 152*, 129–152.
Silveira, M., Alleoni, L., O'Connor, G., & Chang, A. (2006). Heavy metal sequential extraction methods: A modification for tropical soils. *Chemosphere, 64*, 1929–1938.
Texas Commission on Environmental Quality [TCEQ]. (2003). *Surface water quality monitoring procedures: Vol. 1: Physical and chemical monitoring methods for water, sediment and tissue*. Austin, TX: Texas Commission on Environmental Quality.
Texas Natural Resource Conservation Commission [TNRCC]. (1997). *Second phase of the binational study regarding the presence of toxic substances in the Rio Grande/Rio Bravo and its tributaries along the boundary portion between the United States and Mexico* (Final report). Austin, TX: Texas Natural Resource Conservation Commission.
TNRCC. (2002). *State of the Rio Grande and the environment of the border region, strategic plan* (Fiscal years 2003–2007, Vol. 3). Austin, TX: Texas Natural Resource Conservation Commission.
United States Environmental Protection Agency [U.S.EPA]. (1994). *Method 200.8, determination of trace elements in waters and wastes by inductively coupled plasma–mass spectrometry*. Cincinnati, OH: Environmental Monitoring Systems Laboratory, Office of Research and Development.
United States Environmental Protection Agency [U.S.EPA]. (1997). *Damage cases and environmental releases from mines and mineral processing sites*. Washington, DC: Offices of Solid Waste, U.S.EPA.
United States Environmental Protection Agency [U.S.EPA]. (2005). *Guidance for developing ecological soil screening levels*. Washington, DC: Office of Solid Waste and Emergency Response.
United States Environmental Protection Agency [U.S.EPA]. (2007). *Microwave assisted acid digestion of sediments, sludge, soils and oils*. Washington, DC: U.S.EPA.
United States Section, International Boundary and Water Commission [U.S.IBWC]. (1994). *Binational study regarding the presence of toxic substances in the Rio Grande/Rio Bravo and its tributaries along the boundary portion between the United States and Mexico*. El Paso, TX: International Boundary and Water Commission.
Van Geen, A., Takesue, R., & Chase, Z. (1999). Acid mine tailings in southern Spain. *The Science of the Total Environment, 242*, 221–229.
Vaze, J., & Chiew, F. (2002). Experimental study of pollutant accumulation on an urban road surface. *Urban Water, 4*, 379–389.
Wulf, H., Bookhagen, B., & Scherler, D. (2012). Climatic and geologic controls on suspended sediment flux in the Sutlej River Valley, western Himalaya. *Hydrology and Earth System Sciences, 16*(7), 2193–2217. doi:10.5194/hess-16-2193-2012.

Chapter 9
Characterization of Brackish Groundwater Desalination Concentrate Discharge Impacts on Water Quality in a Texas Coastal Area

Yaneth P. Gamboa and Lee W. Clapp

Abstract Management of concentrate streams is one of the major environmental concerns for desalination operations. This study characterized the impacts of concentrate discharge on receiving waters downstream of the Southmost Regional Water Authority (SRWA) brackish groundwater desalination plant, which is located close to the Gulf Coast near Brownsville, Texas. Water quality was characterized with respect to dissolved oxygen (DO), temperature, total dissolved solids (TDS), pH, total and ortho-phosphorus (P), and arsenic (As). The study also generated baseline water quality data as an initial step for evaluating the potential impacts that could occur if the concentrate stream were used as supplement water for the re-flooding and restoration of the Bahia Grande wetland area, which is located within the Laguna Atascosa National Wildlife Refuge. The study showed that the concentrate stream did not appear to have a significant adverse impact on downstream receiving waters and contributed to dilution of salinity in San Martin Lake. However, given projections that brackish groundwater desalination will meet 20 % of the Lower Rio Grande Valley's municipal water supply within the next 50 years, a critical assessment of the potential cumulative impacts of region-wide brackish groundwater desalination on surface water quality in the region is needed.

Keywords Arsenic • Brackish groundwater • Concentrate • Desalination • Phosphorus • Salinity

---

Y.P. Gamboa • L.W. Clapp (✉)
Department of Environmental Engineering, Frank H. Dotterweich College of Engineering, Texas A&M University-Kingsville, Kingsville, TX 78363-8202, USA
e-mail: lee.clapp@tamuk.edu

D. Ramirez et al. (eds.), *Environmental Sustainability Issues in the South Texas–Mexico Border Region*, DOI 10.1007/978-94-007-7122-2_9,

9.1 Introduction

The Lower Rio Grande Valley (LRGV) region in South Texas is undergoing intense population growth, urban sprawl, and significant water resource challenges (Rogers, Sturdivant, Rister, Lacewell, & Harris, 2008). With an existing population of about 1.6 million in 2010, municipal populations in the region are projected to surpass 3.8 million by 2060, with 91 % located in Webb, Hidalgo, and Cameron counties along the U.S.-Mexico border. Consequently, municipalities in the region will demand a significantly greater water supply over the next 50 years, with a projected increase from 291,000 acre-feet (AF) per year in 2010 to almost 671,000 AF per year by 2060. The major municipal water demand centers are Laredo, McAllen-Mission-Edinburg, and Brownsville-Harlingen-San Benito. Advanced municipal water conservation and a significant conversion of irrigation water rights to municipal water rights will only meet part of the increasing municipal water demand (NRS Consulting Engineers, 2006a).

Most of the surface water supply in the Rio Grande Valley region is currently from the Amistad and Falcon International Reservoirs on the Rio Grande River, which are operated by the International Boundary and Water Commission (IBWC) for flood control and water supply (NRS Consulting Engineers, 2006a). Although fresh groundwater resources are limited, the region does have abundant brackish groundwater resources, defined as having between 1,000 and 10,000 mg/L of total dissolved solids (TDS). The Texas Commission on Environmental Quality (TCEQ) has established a standard of 1,000 mg/L TDS for public water supply systems (Texas Commission on Environmental Quality [TCEQ], 2004), whereas the federal standard is 500 mg/L (United States Environmental Protection Agency [U.S. EPA], 1992). The major groundwater source in the region is the Gulf Coast Aquifer, which is estimated to be capable of supplying up to 262,330 AF of groundwater per year between 2010 and 2060, of which about 80 % is brackish (LBG-Guyton and Associates, 2003; NRS Consulting Engineers, 2006b).

The need for new high-quality water supplies, together with decreasing desalination costs due to advances in membrane reverse osmosis (RO) technology, has led to a rapid increase in brackish groundwater desalination in South Texas (Norris, 2005). The Rio Grande Regional Water Planning Group (RGRWPG) projected that brackish groundwater desalination will meet 20.4 % of the region's municipal water supply by 2060, with seawater desalination providing an additional 2.3 % (NRS Consulting Engineers, 2006b). In contrast, due to hydrologic and economic reasons, new surface water supplies were projected to supply only 6.0 % of municipal water supply by 2060. A main reason for brackish groundwater playing such a significant role in the region's future water supply is that it is projected to be less costly than new surface water supplies. As an example, Rogers et al. (2008) performed a comprehensive net present value (NPV) economic study to compare the costs of producing potable water in the LRGV via brackish groundwater desalination and conventional surface-water treatment. Their analysis determined the life-cycle cost of producing water at the Southmost Regional Water Authority (SRWA) Regional

Desalination Plant near Brownsville to be $1.89 per 1,000 gal per year, as compared to $1.99 per 1,000 gal per year for the McAllen Northwest surface-water treatment facility (both in 2006 dollars).

Brackish groundwater desalination is already common in the LRGV. The SRWA desalination plant near Brownsville began producing about 7.5 million gallons per day (MGD) of potable water in 2004 and was designed for expansion to 22 MGD. This plant currently provides about 40 % of the annual water supply for the SRWA, which services over 150,000 people. Other brackish groundwater desalination plants in the region include a 0.25 MGD plant in Rancho Viejo, a 1 MGD plant in La Sara, a 2 MGD plant in Primera, 2 separate 3 MGD plants near Edinburg, and a 2 MGD plant in Donna, for a total production of almost 19 MGD (or 21,300 AF/year). More importantly, brackish groundwater desalination in the Lower Rio Grande Valley region is expected to continue to increase, with an additional 62 MGD (or 70,000 AF/year) of brackish groundwater desalination production already proposed (NRS Consulting Engineers, 2006b).

The primary environmental concern associated with desalination is concentrate disposal (United Nations Environmental Programme [UNEP], 2008). In particular, as the number of desalination plants in a region increase, the cumulative impact of concentrate disposal on receiving waters may become a limiting factor (Nederlof & Hoogendoorn, 2005). RO concentrate (also called retentate or brine) contains the concentrated constituents of the raw feed water plus any chemicals (e.g., anti-scalants) added for pretreatment purposes. The characteristics of the RO concentrate depend on the quality of the feed water, the percentage of the feed water recovered as permeate, and the pretreatment methods used. A typical brackish groundwater membrane desalination plant operating at 75 % recovery will produce 1 gall of concentrate for every 3 gall of potable water produced, with the concentrate having a TDS about four times greater than the feed groundwater. Toxic compounds from the groundwater (e.g., arsenic and hydrogen sulfide) and low dissolved oxygen (DO) levels can also sometimes present problems (Chelme-Ayala, Smith, & Gamal El-Din, 2009).

Possible management strategies for RO concentrate disposal include surface water discharge, deep well injection, sewer disposal, evaporation ponds, spray irrigation, and zero liquid discharge using thermal evaporators or crystallizers (Chelme-Ayala et al., 2009). Generally, only surface water discharge or deep well injection is feasible for larger RO desalination plants in Texas (Mickley, 2006). The largest desalination plant in Texas, the 27.5 MGD Kay Bailey Hutchison Desalination Plant near El Paso, uses deep well injection to dispose of the RO concentrate (Hutchison, 2009). However, in the Rio Grande Valley region surface discharge to surface water will probably continue to be the most common approach for RO concentrate management due to the significantly lower costs involved. Some facilities may dilute the concentrate with wastewater plant effluent or cooling water to minimize increases in salinity at the discharge point compared with upstream receiving water (Mickley).

As desalination becomes an increasingly larger component of the water supply portfolio in South Texas, an improved understanding of the potential environmental impacts of RO concentrate disposal is needed. The primary objective of this study was to assess the impacts of concentrate discharge from an existing brackish groundwater desalination plant in South Texas on the receiving surface waters, using the Southmost Regional Water Authority (SRWA) Regional Desalination Plant near Brownsville as a case study. A secondary objective was to generate baseline water quality data as an initial step for evaluating the potential impacts that could occur if the SRWA desalination concentrate discharge were used as supplement water for the re-flooding and restoration of the 6,500-acre Bahia Grande saltwater wetland within the Laguna Atascosa National Wildlife Refuge.

9.2 Site Description

The SRWA Regional Desalination Plant is located near Brownsville in Cameron County, Texas, and has been in operation since 2004. The brackish groundwater (with a TDS of approximately 3,500 mg/L) is continuously pumped at a maximum rate of 9.5 MGD from 20 surrounding wells at depths of 280–300 ft. Until February 2006, pretreatment of the groundwater consisted of 5-μm cartridge filters and the addition of an organo-phosphonate anti-scalant at a dose of 5 ppm. However, beginning in March 2006, an additional pretreatment step was implemented that involved adding chlorine to the feed groundwater at about 0.1 mg/L to oxidize and remove arsenic as a means of complying with the new drinking water standard of 10 μg/L. The average concentration of arsenic in the feed groundwater is about 27 μg/L. After pretreatment, a maximum of 8.0 MGD of the feedwater is pumped at high pressure (180 psi) through 1,386 RO membrane elements arranged in a two-stage design. The plant operates at 75 % recovery, resulting in a maximum permeate stream of 6.0 MGD at a TDS of 40–50 mg/L and a concentrate stream of about 2.0 MGD at a TDS of about 8,750 mg/L. A maximum of 1.5 MGD of the pretreated feedstream bypasses the RO process and is subsequently blended with the 6.0 MGD permeate, resulting in a maximum total potable water production of 7.5 MGD at a TDS of about 300–475 mg/L (Gomez, 2005; Norris, 2004; Sturdivant et al., 2009).

At the time of this study, the plant was discharging a maximum of 2.0 MGD of RO concentrate through a 16-in. diameter pipe into a nearby earthen drainage ditch. If the plant expands to 22.0 MGD, as is planned, concentrate flow will increase to 5.9 MGD. According to the plant's Texas Pollutant Discharge Elimination System (TPDES) discharge permit, the current allowable maximum daily concentrate flow is 4.0 MGD, the allowable pH is between 6 and 9, and the maximum allowable TDS is 12,720 mg/L (Gomez, 2005). The drainage ditch that receives the concentrate stream flows into San Martin Lake, a tidal-influenced estuary that also receives about 5 MGD of treated wastewater effluent from Brownsville's North Wastewater Treatment Plant. San Martin Lake subsequently drains into the Brownsville Ship Channel, which in turn is connected to the Lower Laguna Madre. The Lower

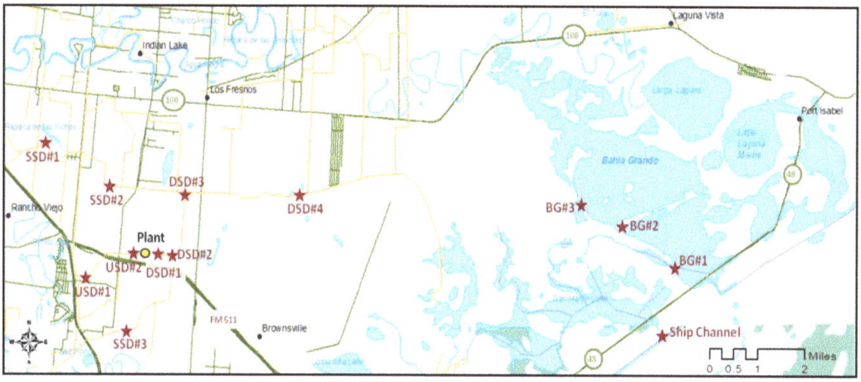

Fig. 9.1 Sampling site locations during summer 2006 (Modified from U.S. Geological Society, 2002)

Laguna Madre is a long, shallow hypersaline lagoon lying between Padre Island and the southern tip of the Texas mainland.

While this study was being conducted, the largest saltwater wetland restoration project in the U.S. was initiated for the Bahia Grande, which is located within the nearby Laguna Atascosa National Wildlife Refuge (U.S. Fish and Wildlife Service, 2003). Before 1936 the Bahia Grande was a large tidal-influenced lagoon. In 1936, the Port of Brownsville constructed the Brownsville Ship Channel, cutting off tidal flow to the Bahia from the Lower Laguna Madre. After losing tidal inflow the Bahia Grande became a salty, arid area that only received water from rainfall and created dust nuisances for surrounding areas. In July 2005, restoration started with the construction of a pilot channel between the Bahia and the Brownsville Ship Channel (Fig. 9.1) and the re-flooding of approximately 6,500 acres.

9.3 Methods

9.3.1 Sampling Methods

Water sampling for analysis of DO, temperature, conductivity, pH, dissolved total phosphorus and ortho-phosphorus, and dissolved total arsenic was conducted in summer 2006 and again in spring 2007 (Gamboa, 2007). Surface water sampling followed procedures developed by the Texas Commission on Environmental Quality (TCEQ, 2003). Five sampling events took place in summer 2006 (June 8, July 10, July 24, July 31, and August 28) and five in spring of 2007 (March 10, March 17, March 23, March 31, and May 1). For the first sampling phase, grab samples were collected at 14 sites (Fig. 9.1), which included the plant outfall (Plant), two upstream ditch (USD) sites, four downstream ditch (DSD) sites, three side-stream ditch (SSD) sites, three Bahia Grande (BG) sites, and a Brownsville Ship Channel (Ship Channel)

Fig. 9.2 Sampling site locations during spring 2007 (Modified from U.S. Geological Society, 2002)

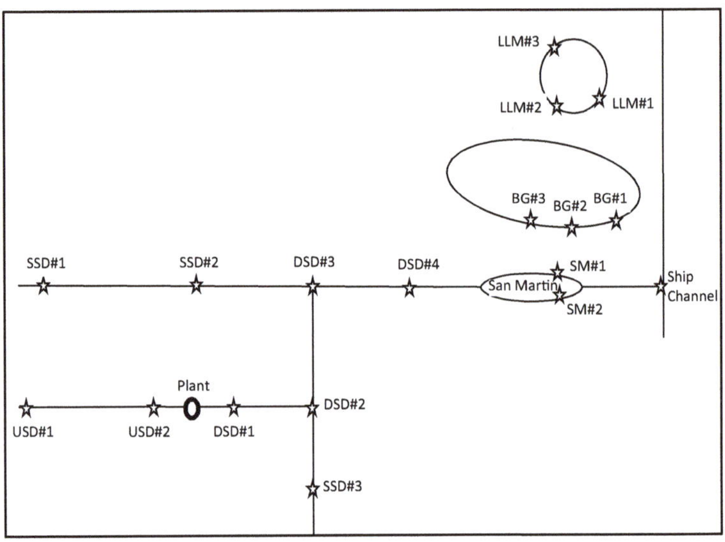

Fig. 9.3 Schematic map of sampling sites, including the plant discharge site, two upstream ditch (*USD*) sites, four downstream ditch (*DSD*) sites, three side-stream ditch (*SSD*) sites, two sites in San Martin Lake (*SM*), three sites in the Bahia Grande (*BG*), three sites in the Little Laguna Madre (*LLM*), and one site in the Brownsville Ship Channel

site near the San Martin Lake drainage point. For the second sampling phase, grab samples were collected from 12 sampling sites (Fig. 9.2) including seven of the previous sites, two sites at San Martin Lake (SM), and three at the Little Laguna Madre (LLM). Water flowed from east to west in all of the sampled drainage ditches. Figure 9.3 shows the relative locations of all the sampling points.

Grab water samples were collected at each sampling point, transferred to 1-L plastic sample bottles with no headspace, and placed inside a cooler. After arriving at the laboratory, the water samples were immediately passed through 0.45-μm membrane filters. The samples to be analyzed for dissolved arsenic were also acidified with 2 mL of nitric acid per liter of sample. All samples were placed in amber plastic bottles and stored in a refrigerator until analyzed.

9.3.2 Analytical Methods

9.3.2.1 Dissolved Oxygen and Temperature

Dissolved oxygen (DO) and temperature were measured at each sampling site using a pre-calibrated YSI Model 85 m.

9.3.2.2 Specific Conductivity, TDS, and pH

Specific conductivity and pH were measured immediately after arriving at the laboratory (before filtering the samples) using a calibrated Thermo Orion 4-Star meter. The pH probe was calibrated with two standard solutions at pH of 7 and 10. The conductivity probe was calibrated using a standard solution of 1,413 μS/cm at 25 °C. Total dissolved solids (TDS) concentrations were approximated by multiplying the conductivity values by an empirical correlation factor of 0.65 (TCEQ, 2010b).

9.3.2.3 Ortho-Phosphorus

Dissolved ortho-phosphorus concentrations were determined by the *Standard Methods*' 4500-P D stannous chloride reactive phosphate method (Clesceri, Greenberg, & Eaton, 1998) using a spectrophotometer set at 690 nm. Six phosphate calibration standards were prepared at concentrations of 0, 40, 80, 120, 160, and 200 ppb (as phosphorus) by diluting a 50 ppm stock standard with Type I deionized (DI) water. Calibration standards and samples were analyzed following the same procedure, and each sample was analyzed in triplicate.

9.3.2.4 Total Phosphorus

In summer 2006, dissolved total phosphorus concentrations were determined using inductively coupled plasma-mass spectrometry (ICP-MS) following the same procedure as described below for arsenic, with the exception that calibration standards were prepared as described for ortho-phosphorus. Calibration curves for total phosphorus were prepared by carrying a series of ortho-phosphorus standards as previously stated through the persulfate digestion step. In spring 2007, dissolved total phosphorus concentrations were determined by the *Standard Methods*' 4500-P (B)(5) persulfate digestion sample preparation method (to convert organic-bound phosphorus to ortho-phosphorus) and followed by the stannous chloride reactive phosphate method. Samples and calibration standards of 50 mL were heated for 30 min in an autoclave at 121 °C after adding sulfuric acid solution and ammonium persulfate. After autoclaving, the samples were allowed to cool to room temperature and then neutralized by adding 1 N sodium hydroxide solution. DI water was added to each sample to a total volume of 100 mL, and these 100-mL samples were analyzed by the stannous chloride reactive phosphate method.

9.3.2.5 Arsenic

To determine dissolved total arsenic concentrations, samples were analyzed by ICP-MS using EPA Method 200.8 (Creed, Brockhoff, & Martin, 1994) with a Thermo Electron X Series instrument. Since arsenic is difficult to measure at sub-ppb levels due to argon-based interferences (ArCl), the collision cell technology (CCT) performance option was used to reduce these interferences (Thermo Electron, 2006). Samples, calibration standards, laboratory fortified blanks, and quality control samples were prepared with a solution of 2 % nitric acid (Omni Trace™) and Type I DI water from a Barnstead NanoPure Diamond™ system. Indium (In) was added to all of the samples as an internal standard to correct for instrument drift and interferences. Due to the high TDS content, the water samples were diluted at a volume proportion of 1-to-10. Calibration curves were prepared using four calibration standards with arsenic concentrations of 0, 1, 10, and 100 ppb; for each calibration standard a minimum of three replicate analyses were performed.

9.3.2.6 Quality Control

Standard Methods quality control procedures were followed and are described in detail by Gamboa (2007).

9.4 Results

9.4.1 Dissolved Oxygen and Temperature

Figure 9.4 shows the average DO and temperature values at sites sampled during summer 2006. The average temperature of the plant discharge was 28 °C, which was 4–7 °C cooler than at the two upstream ditch (USD) sampling locations. The average DO in the discharge was 1.75 mg/L, significantly lower than at all the other sampling locations. This did not result in low DO concentrations in the drainage ditch, however, as the average value at the first downstream ditch site (DSD#1) was 6.60 mg/L. The absence of a more significant drop in DO in the drainage ditch can

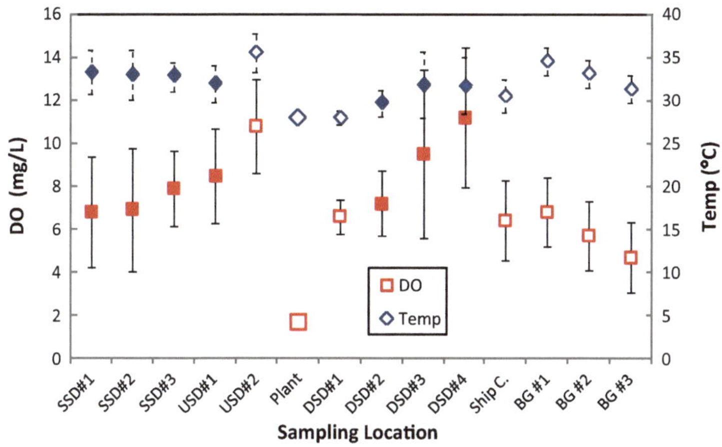

Fig. 9.4 Average DO concentrations and temperatures for sites sampled in summer 2006. Locations with *open symbols* were also sampled the following spring. *Error bars* are 95 % confidence intervals

probably be attributed to the discharge plunging about 6 ft into the drainage ditch, which promoted oxygen transfer. Finally, the average DO values slightly exceeded saturation at four sampling locations: two upstream of the plant at sites USD#1 and USD#2 and two downstream at sites DSD#3 and DSD#4. It is possible that supersaturation occurred due to photosynthesis associated with visibly high algae concentrations.

Figure 9.5 shows the average DO and temperature values for sites sampled during spring 2007. The average temperature of the concentrate discharge was 26 °C. The lowest average DO was again in the plant discharge (3.37 mg/L), which was significantly lower than the values at all other sampling locations. This again did not significantly decrease DO in the drainage ditch, as the average concentration measured at the downstream sampling site DSD#1 was 5.82 mg/L.

9.4.2 *Total Dissolved Solids and pH*

Figure 9.6 shows average TDS and pH values measured for sites sampled during summer 2006. The TDS concentrations were approximated by multiplying the measured conductivity values (in units of μS/cm) by an empirical correlation factor of 0.65 (TCEQ, 2010b). The average TDS concentration in the plant discharge was 8,848 mg/L, as compared to 1,345 and 2,602 mg/L at the two USD sites and 8,049 and 7,387 mg/L at the two nearest DSD sites. The fact that the downstream concentrations were almost as high as the concentrate TDS was consistent with the discharge constituting most of the drainage ditch flow during dry weather. The plant discharge was correlated with an average TDS increase of 5,547 mg/L in the

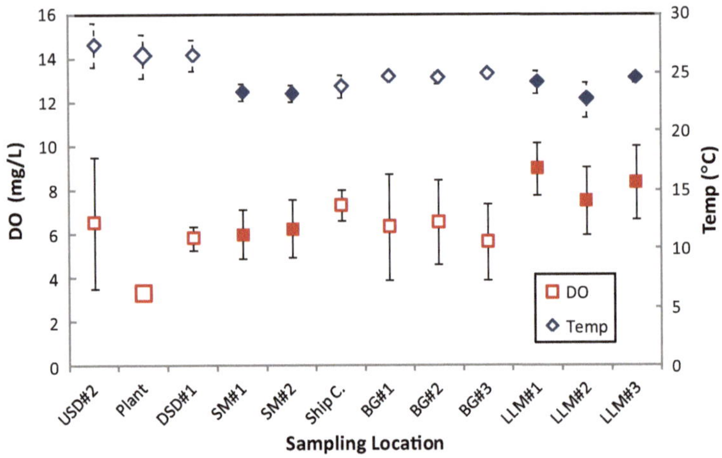

Fig. 9.5 Average DO concentrations and temperatures for sites sampled in spring 2007. Locations with *open symbols* were also sampled the previous summer. *Error bars* are 95 % confidence intervals

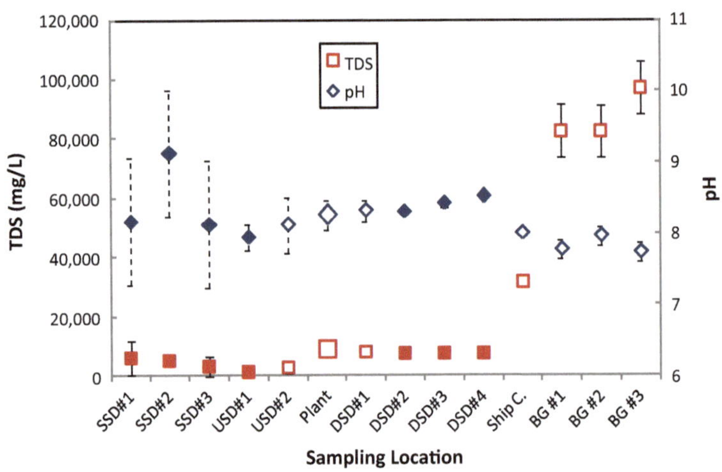

Fig. 9.6 Average TDS and pH values for sites sampled in summer of 2006. Locations with *open symbols* were also sampled the following spring. *Error bars* are 95 % confidence intervals

drainage ditch. However, the average downstream TDS concentrations were not much higher than at three unaffected SSD sites, which had average values ranging from 3,023 to 6,011 mg/L. The average TDS concentration in the Brownsville Ship Channel, into which the plant discharge ultimately flowed, was 31,698 mg/L or 3.6 times greater than for the plant concentrate. The average TDS concentrations at the three Bahia Grande (BG) sampling sites ranged from 82,420 to 97,067 mg/L or at least 9.3 times greater than for the plant concentrate. The very high TDS concentrations in Bahia Grande can be attributed to the fact that it had just recently been re-flooded. The average pH of the plant discharge was 8.3, which fell within the range of 7.9–9.1 observed for all

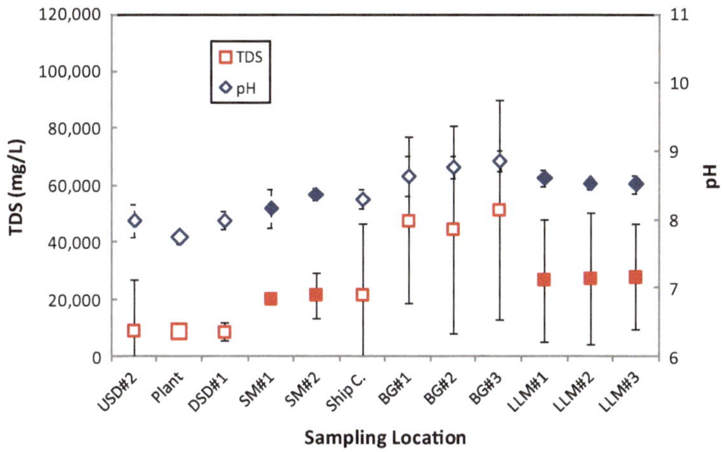

Fig. 9.7 Average TDS and pH values for sites sampled in spring of 2007. Locations with *open symbols* were also sampled the previous summer. *Error bars* are 95 % confidence intervals

the other sampling sites. Finally, the high variability observed for pH at some of the drainage ditch sampling locations may have been due to diurnal photosynthesis carried out by visibly high algae concentrations.

Figure 9.7 shows average TDS and pH values at sites sampled during spring 2007. The average TDS in the plant discharge was 8,702 mg/L, which was within 2 % of the TDS recorded the previous summer. In comparison, the average TDS at the nearest upstream site (USD#2) was over three times higher than the previous summer (8,814 versus 2,602 mg/L); the reason for the significant increase was not known. The average TDS at the nearest downstream ditch site (8,047 mg/L) was almost the same as for the plant discharge, which was consistent with the discharge constituting most of the flow during dry weather. The TDS at the two San Martin Lake (SML) sites, Ship Channel site, and three Little Laguna Madre (LLM) sites all ranged from 19,982 to 27,944 mg/L, consistent with their being tidal-influenced. The average TDS at the three Bahia Grande (BG) sites ranged from 44,541–51,388 mg/L, or roughly half the values observed the previous summer of 82,420–97,067 mg/L. This decrease could be attributed to dilution by the continual inflow of seawater from the Ship Channel since the water level in the Bahia Grande was observed to have increased significantly between summer 2006 and spring 2007. Finally, the average pH of the plant discharge was 7.8, which was slightly lower than the range of 8.0–8.9 observed for all the other sampling locations.

9.4.3 *Total and Ortho-Phosphorus*

Figure 9.8 shows average dissolved total phosphorus concentrations for sites sampled during summer 2006. The average total phosphorus concentration in the

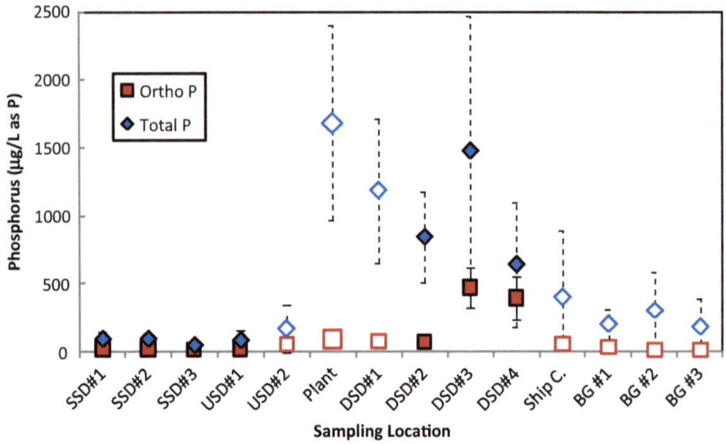

Fig. 9.8 Average dissolved total and ortho-phosphorus at sites sampled in summer 2006. Locations with *open symbols* were also sampled the following spring. *Error bars* are 95 % confidence intervals

plant discharge was 1,682 µg/L as compared to 82 and 164 µg/L at the two USD sampling sites and 1,183 and 842 µg/L at the first two DSD sampling sites. The plant discharge was thus correlated with an average total phosphorus increase of 1,019 µg/L in the receiving drainage ditch. The elevated total phosphorus concentrations in the plant discharge were primarily due to the use of a proprietary organo-phosphonate anti-scalant as part of the RO pretreatment process. All downstream ditch total phosphorus concentrations were significantly higher (ranging from 640 to 1,479 µg/L) than at the three SSD sites (from 45 to 92 µg/L) and USD (from 82 to 164 µg/L). Finally, the average dissolved total phosphorus concentration in the Brownsville Ship Channel was 396 µg/L and the average concentrations at the three Bahia Grande (BG) sampling sites ranged from 176 to 303 µg/L.

Figure 9.8 also shows average dissolved ortho-phosphorus concentrations for sites sampled during summer 2006. The average ortho-phosphorus concentration in the plant discharge was 87 µg/L as compared to 19 and 48 µg/L at the two USD sites and 72 and 71 µg/L at the first two downstream ditch sites (DSD#1 and DSD#2). The plant discharge was correlated with an average ortho-phosphorus increase of only 24 µg/L between the nearest upstream and downstream sampling sites and was not statistically significant. However, the ortho-phosphorus concentrations at sites DSD#3 and DSD#4 increased to 469 and 390 µg/L, respectively, which were significantly higher than at the two upstream ditch sites and three side-stream ditch sites. Although the reason for the significant increase in ortho-phosphorus concentrations (from 71 to 469 µg/L) between the DSD#2 and DSD#3 locations was not known, the observed trend was consistent with the possibility that hydrolysis, photolysis, and/or biodegradation had slowly transformed the organo-phosphonate discharged with the concentrate to ortho-phosphorus. Finally, the average ortho-phosphorus concentration in the Brownsville Ship Channel was

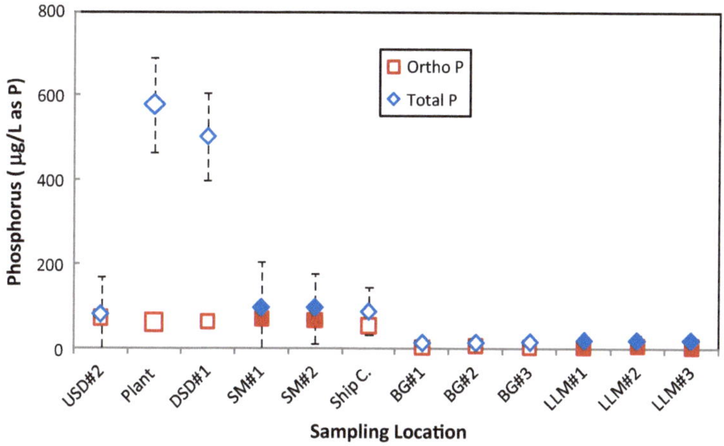

Fig. 9.9 Average dissolved total and ortho-phosphorus at sites sampled in spring 2007. Locations with *open symbols* were also sampled the previous summer. *Error bars* are 95 % confidence intervals

54 µg/L, while the average concentrations at the three Bahia Grande (BG) sampling sites ranged from 7 to 26 µg/L.

Figure 9.9 shows average dissolved total phosphorus concentrations for sites sampled during spring 2007. The average total phosphorus concentration in the plant discharge was 580 µg/L as compared to 82 µg/L at the nearest upstream sampling site (USD#2) and 503 µg/L at the nearest downstream sampling site (DSD#1). The plant discharge was correlated with an average total phosphorus increase of 421 µg/L in the receiving ditch that again could be attributed to the use of an organo-phosphonate anti-scalant in the RO pretreatment process. However, the average discharge total phosphorus concentration in spring 2007 (580 µg/L) was only 35 % of the average concentration in summer 2006 (1,682 µg/L). The reason for this decrease was unknown as there was no reported change in the anti-scalant dosing. The average total phosphorus concentration at the Brownsville Ship Channel site was 89 µg/L compared to 396 µg/L the previous summer, and the average concentrations at the three Bahia Grande (BG) sampling sites ranged from 15 to 17 µg/L compared to 176 to 303 µg/L the previous summer. The significant decrease at the Bahia Grande sites could be attributed to dilution by the continual inflow of seawater from the Brownsville Ship Channel. The average total phosphorus concentration for the three Little Laguna Madre (LLM) sampling sites was 19 µg/L.

Figure 9.9 also shows average dissolved ortho-phosphorus values for sites sampled during spring 2007. The average ortho-phosphorus concentration in the plant discharge was 63 µg/L compared to 73 µg/L at the nearest upstream ditch site (USD#2) and 65 µg/L at the nearest downstream ditch site (DSD#1). Thus, the plant discharge was not correlated with any significant change in ortho-phosphorus between the nearest upstream and downstream sampling sites. The spring 2007 results did not provide additional data to assess the possibility that the organo-phosphonate discharged with

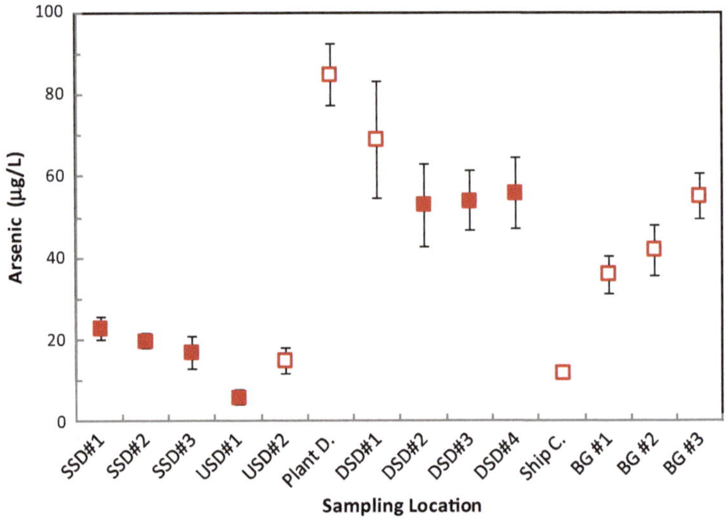

Fig. 9.10 Average dissolved arsenic concentrations at sites sampled in summer 2006. Locations with *open symbols* were also sampled the following spring. *Error bars* are 95 % confidence intervals

the concentrate slowly transformed to ortho-phosphorus in the downstream drainage ditch (consistent with the summer 2006 data), because sites DSD#2, DSD#3, and DSD#4 were not re-sampled. The average discharge ortho-phosphorus concentration in spring 2007 (63 µg/L) was 72 % of that observed in summer 2006 (87 µg/L). Finally, the ortho-phosphorus concentrations at the two San Martin Lake and Ship Channel sites ranged from 55 to 70 µg/L, and the concentrations at the three Bahia Grande and three Little Laguna Madre (LLM) sites ranged from 4 to 8 µg/L.

9.4.4 Dissolved Arsenic

Figure 9.10 shows average dissolved arsenic concentrations for sites sampled during summer 2006. The average arsenic concentration in the plant discharge was 85 µg/L compared to 6 and 15 µg/L at the two USD sites and 69 and 53 µg/L at the first two DSD sites. The plant discharge was correlated with an average arsenic increase of 54 µg/L in the receiving ditch. The arsenic in the plant discharge was due to the presence of between 10 and 80 µg/L in the 20 wells used to supply groundwater to the plant (J. Garza, 2007, November 26, personal interview). All the downstream ditch arsenic concentrations (between 53 and 69 µg/L) were significantly higher than at the three SSD sites (from 17 to 23 µg/L) and upstream ditch sites (from 6 to 15 µg/L). Finally, the average arsenic concentration in the Ship Channel was 12 µg/L, and the average concentrations at the three Bahia Grande sampling sites ranged from 36 to 55 µg/L.

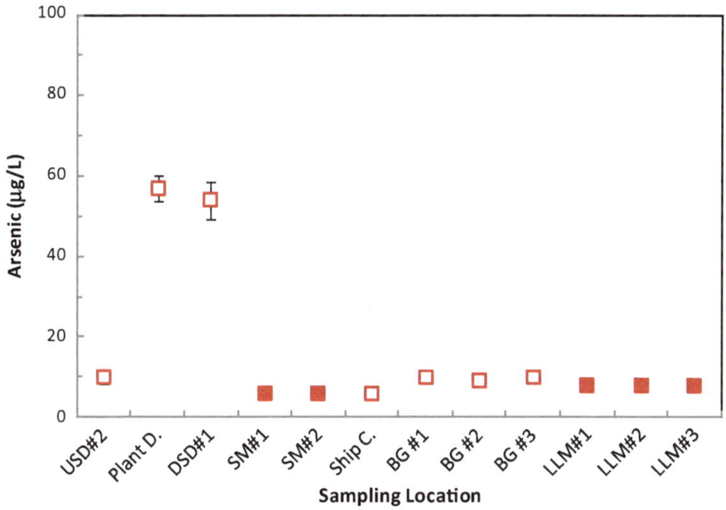

Fig. 9.11 Average dissolved arsenic concentrations at sites sampled in spring 2007. Locations with *open symbols* were also sampled the previous summer. *Error bars* are 95 % confidence intervals

Figure 9.11 shows average dissolved arsenic concentrations for sites sampled during spring 2007. The average arsenic concentration in the plant discharge was 57 μg/L compared to 10 μg/L at the nearest upstream sampling site (USD#2) and 54 μg/L at the nearest downstream sampling sites (DSD#1). The plant discharge was thus correlated with an average arsenic increase of 44 μg/L in the receiving ditch that could be attributed to the presence of arsenic at 10 to 80 μg/L in the 20 wells used to supply groundwater to the plant. The discharge average arsenic concentration in spring 2007 (57 μg/L) was 67 % of that for summer 2006 (85 μg/L). The average arsenic concentration at the Brownsville Ship Channel site was 6 μg/L compared to 12 μg/L the previous summer, and the average concentrations at the three Bahia Grande sampling sites ranged from 9 to 11 μg/L compared to 36 to 55 μg/L the previous summer. The decreased concentrations at the Bahia Grande sites could again be attributable to dilution by the continual inflow of seawater from the Ship Channel. The average dissolved arsenic concentration for the three LLM sampling sites was 8 μg/L.

9.5 Discussion

9.5.1 Dissolved Oxygen and Temperature

Drainage ditches that flow into the Brownsville Ship Channel are designated by the TCEQ as having a "limited aquatic life" use and correspondingly have a minimum

DO criterion of 3 mg/L (TCEQ, 2010a). Although the plant concentrate had average DO concentrations of 1.75 mg/L (summer 2006) and 3.37 mg/L (spring 2007), this did not appear to adversely impact the DO in the receiving drainage ditch since none of the sampled downstream sites had average DO values below 5.8 mg/L. The DO in the drainage ditches did vary significantly possibly due to diurnal fluctuations associated with photosynthesis carried out by visibly high algae concentrations. Finally, the concentrate discharge cooled the water in the drainage ditch by about 5 °C during summer 2006 but had no significant impact during spring 2007.

9.5.2 Total Dissolved Solids and pH

The average TDS concentrations in the plant concentrate (estimated from conductivity measurements) in summer 2006 and spring 2007 were 8,848 and 8,702 mg/L, respectively, both well below the plant's TPDES permit limit of 12,720 mg/L. In comparison, the TDS in surrounding drainage ditches unaffected by the plant discharge ranged from 1,350 to 6,000 mg/L. The average TDS in the plant concentrate was less than one-half of that in San Martin Lake and the Brownsville Ship Channel and less than one-fifth of that in the Bahia Grande. The desalination plant concentrate actually contributed to dilution of TDS in San Martin Lake thus supporting the concept of using concentrate from brackish groundwater desalination plants as a water supply for estuarine wetlands (Chelme-Ayala et al., 2009). Finally, the average pH values of the plant concentrate in summer 2006 and spring 2007 were not significantly different from those found in surrounding drainage ditches.

9.5.3 Total and Ortho-Phosphorus

The SRWA desalination plant's TPDES discharge permit did not include a limit for phosphorus. At the time this study was performed, the TCEQ had not established numerical ambient water quality criteria for phosphorus to which the measured dissolved total and ortho-phosphorus concentrations could be compared. However, the revised *Procedures to Implement the Texas Surface Water Quality Standards* (TCEQ, 2010b) indicate that when reductions in phosphorus loading are required to protect surface water quality, daily average effluent limits for total phosphorus will range from 500 μg/L (for permitted discharges of >3.0 MGD) to 1,000 μg/L (for permitted discharges of <0.5 MDG). In comparison, the SRWA plant's approximately 2 MGD concentrate stream had average total phosphorus concentrations of 1,682 μg/L (summer 2006) and 580 μg/L (spring 2007). In addition, in a ruling with national implications (MacCurdy, 2010), the U.S. EPA recently proposed instream total phosphorus standards for Florida, depending on the geographic location of the watershed that ranges from 43 to 749 μg/L (Federal Register, 2010). In comparison, the SRWA plant concentrate discharge was correlated with average total phosphorus increases in the receiving ditch of 1,019 μg/L (summer of 2006) and 421 μg/L (spring 2007). The SRWA plant did not discharge its concentrate stream into receiving waters categorized as impaired by TCEQ and was unlikely to have its discharge permit amended to include total phosphorus limits in the near future. Future brackish groundwater desalination plants in South Texas discharging into the Arroyo Colorado, classified as an impaired water body (TCEQ, 2010c), could be subjected to total phosphorus permit limits that would restrict the use of phosphonate-based anti-scalants.

In contrast to the total phosphorus results, the SRWA plant's concentrate discharge was not correlated with significant increases in ortho-phosphorus between the nearest upstream and downstream sampling sites, but was correlated with significantly higher ortho-phosphorus concentrations at sampling locations further downstream (DSD#3 and DSD#4). These results were consistent with the possibility that the organo-phosphonate anti-scalant present in the discharged concentrate slowly transformed to ortho-phosphorus in the drainage ditch via hydrolysis, photolysis, and/or biodegradation. Whereas eutrophication problems have been observed near the outlets of desalination plants in which polyphosphate anti-scalants were used, as they easily hydrolyze to ortho-phosphate, phosphonates are generally considered to be fairly stable with low biodegradation rates (Lattemann & Höpner, 2008), even though they are considered "inherently biodegradable" (UNEP, 2008, page 93). Unfortunately, little is known about the fate of phosphonate anti-scalants in the environment. A recent United Nations Environmental Programme assessment concluded that "no field investigations about the actual environmental fate and interactions of anti-scalants have been carried out to date" (UNEP, page 94).

9.5.4 Arsenic

The average dissolved arsenic concentrations in the plant concentrate in summer 2006 and spring 2007 were 85 and 57 μg/L, respectively, due to the presence of arsenic in the feed groundwater and a four-fold concentration in the RO process. In comparison, the dissolved arsenic in the drainage ditches unaffected by the plant discharge ranged from 6 to 23 μg/L. The plant discharge was correlated with average arsenic increases in the receiving ditch of 54 μg/L (summer 2006) and 44 μg/L (spring 2007). Moreover, the average arsenic concentration in the concentrate discharge was roughly ten times greater than that in San Martin Lake, the Brownsville Ship Channel, the Little Laguna Madre, and the Bahia Grande (in spring 2007 after the initial re-flooding phase). Nonetheless, dissolved arsenic concentrations within the receiving drainage ditch never exceeded the TCEQ's acute or chronic arsenic ambient water quality criteria of 340 and 150 μg/L, respectively, established to protect aquatic life in freshwater surface bodies

(TCEQ, 2010a). In addition, arsenic concentrations in San Martin Lake, the Brownsville Ship Channel, the Bahia Grande, and the Little Laguna Madre never exceeded TCEQ's acute or chronic arsenic ambient water quality criteria of 149 and 78 μg/L, respectively, for saltwater surface bodies.

9.6 Conclusions

Being located near the coast, the SRWA brackish groundwater desalination plant has the benefit of a simple and low-cost concentrate disposal system. This study showed that the concentrate discharge did not significantly degrade water quality in the receiving ditch or downstream water bodies (i.e., San Martin Lake and the Brownsville Ship Channel) with respect to DO, temperature, TDS, pH, total phosphorus, ortho-phosphorus, and arsenic. Moreover, the concentrate discharge actually contributed to dilution of TDS in San Martin Lake, and also had much lower TDS concentrations than in the recently restored Bahia Grande. These findings were consistent with previous studies that have used concentrate streams as water supply for estuarine wetlands (Frank, Bays, & Ortega, 2007).

However, the study results cannot be directly extrapolated to all future brackish groundwater desalination plants in the LRGV because facilities that do not have direct access to the coast may have to discharge concentrate streams to inland receiving waters that have impaired quality (e.g., the Arroyo Colorado). Given projections that brackish groundwater desalination in the region will meet 20 % of the LRGV's municipal water supply within the next 50 years (NRS Consulting Engineers, 2006b), a critical assessment of the potential cumulative impacts of concentrate disposal on regional surface water quality in the LRGV is needed.

Acknowledgments This material is based upon work supported by the Center of Research Excellence in Science and Technology – Research on Environmental Sustainability of Semi-Arid Coastal Areas (CREST-RESSACA) at Texas A&M University-Kingsville (TAMUK) through a Cooperative Agreement (No. HRD-0734850) from the National Science Foundation (NSF). Any opinions, findings, and conclusions or recommendations expressed in this material are those of the authors and do not necessarily reflect the views of the National Science Foundation. The authors acknowledge the assistance and cooperation of the following people during the sampling periods of this study: Dr. Jude Benavides and Clinton Roberts (University of Texas at Brownsville), Linda L. Miller (Deputy Refuge Manager, Laguna Atascosa National Wildlife Refuge), and Jose Garza (Chief Operator, SRWA Desalination Plant).

References

Chelme-Ayala, P., Smith, D., & Gamal El-Din, M. (2009). Membrane concentrate management options: A comprehensive critical review. *Canadian Journal of Civil Engineering, 36*, 1107–1119.

Clesceri, L., Greenberg, A., & Eaton, A. (Eds.). (1998). *Standard methods for examination of water & wastewater* (20th ed.). Washington, DC: American Public Health Association.

Creed, J., Brockhoff, C., & Martin, T. (1994). *Method 2008 determination of trace elements in waters and wastes by inductively coupled plasma-mass spectrometry*. Cincinnati, OH: Environmental Monitoring Systems Laboratory, Office of Research and Development, U.S. Environmental Protection Agency.

Federal Register. (2010, January 26). *Water quality standards for the State of Florida* (Proposed rule, 75(16), p. 4174). Retrieved July 26, 2010, from www.epa.gov/waterscience/standards/rules/florida/

Frank, B., Bays, J., & Ortega, K. (2007, October 15–17). Natural treatment of membrane concentrate for beneficial use in Oxnard, California. In *The Proceedings of the Water Environment Federation*. San Diego, CA: WEFTEC.

Gamboa, Y. (2007). *Characterization of desalination plant concentrate discharge impacts on water quality in a Texas coastal area*. M.S. dissertation, Texas A&M University – Kingsville, Kingsville. Retrieved July 22, 2010, from Dissertations & Theses @ Texas A&M System, (Publication No. AAT 1445784).

Gomez, G. (2005, October 20). *Design and construction of a 7.5 MGD desalination plant for the city of Brownsville*. Paper presented at the Emerging Technologies for a Sustainable Environment Conference, South Padre Island, TX.

Hutchison, W. R. (2009). Desalination of brackish groundwater and deep well injection of concentrate in El Paso, Texas. In *The Proceedings of World Environmental and Water Resources Congress 2009*, Kansas City, MO. Retrieved July 22, 2010, from http://dx.doi.org/10.1061/41036(342)551. doi:10.1061/%2041036(342)551

Lattemann, S., & Höpner, T. (2008). Environmental impact and impact assessment of seawater desalination. *Desalination, 220*, 1–15.

LBG-Guyton and Associates. (2003). *Brackish groundwater manual for Texas regional water planning groups*. Prepared for the Texas Water Development Board in association with NRS Consulting Engineers. Retrieved January 10, 2010, from http://www.twdb.state.tx.us/publications/reports/contracted_reports/doc/2001483395.pdf

MacCurdy, M. (2010, February 3). EPA proposal for numeric nutrient standards has national implications. *Martin Law Newsletter*. Retrieved July 26, 2010, from www.martenlaw.com/newsletter/20100203-numeric-nutrient-standards

Mickley, M. (2006). *Membrane concentrate disposal: Practices and regulation*, (2nd ed.). Denver, CO: U.S. Department of the Interior, Bureau of Reclamation, Technical Service Center, Water Treatment Engineering and Research Group. Retrieved January 10, 2010, from http://www.usbr.gov/research/AWT/reportpdfs/report123.pdf

Nederlof, M., & Hoogendoorn, J. (2005). Desalination of brackish groundwater: The concentrate dilemma. *Desalination, 182*, 441–447.

Norris, J. (2004). The southmost regional water authority desalination plant. In *The future of desalination in Texas* (Vol. 2). Austin, TX: Texas Water Development Board. Retrieved February 15, 2010, from http://www.twdb.state.tx.us/publications/reports/numbered_reports/doc/R363/D7.pdf

Norris, J. (2005). Brackish groundwater desalination in South Texas: An alternative to the Rio Grande. In R. Mace, S. Davidson, E. Angle, & W. Mullican (Eds.), *Aquifers of the gulf coast of Texas* (Texas Water Development Board Report 365). Austin, TX: Texas Water Development Board. Retrieved January 10, 2010, from www.twdb.state.tx.us/publications/reports/GroundWaterReports/GWReports/R365/AGCindex.htm.

NRS Consulting Engineers. (2006a). *Rio Grande regional water planning group region M regional water plan, final plan, chapter 2: Current and projected population and water demand for the Rio Grande region*. Revised in June, 2010. Retrieved July 22, 2010, from www.riograndewaterplan.org/downloads/Chapter_2_Final_Plan_Revised_2010.pdf

NRS Consulting Engineers. (2006b). *Rio Grande regional water planning group region M regional rater plan, final plan, chapter 4: Identification, evaluation, & selection of water management strategies based on need*. Revised June, 2010. Retrieved July 22, 2010, from www.riograndewaterplan.org/downloads/Chapter_4_Final_Plan_Revised_2010.pdf

Rogers, C., Sturdivant, A., Rister, M., Lacewell, R., & Harris, B. (2008, February 2–6). *Economic implications of conventional water treatment versus desalination: A dual case study*. Paper presented at the Southern Agricultural Economics Association Annual Meeting, Dallas, TX.

Sturdivant, A., Rister, M., Rogers, C., Lacewell, R. Norris, J., Leal, J., et al. (2009). *An analysis of the economic and financial life-cycle costs of reverse osmosis desalination in South Texas: A case study of the southmost facility* (Texas Water Resources Institute Report TR-295). Retrieved July 10, 2010, from http://twri.tamu.edu/reports/2009/tr295.pdf

Texas Commission on Environmental Quality [TCEQ]. (2003). *Surface water quality monitoring procedures, Vol. 1: Physical and chemical monitoring methods for water, sediment, and tissue* (RG-415). Retrieved July 22, 2010, from http://www.tceq.texas.gov/publications/rg/rg-415

Texas Commission on Environmental Quality [TCEQ]. (2004). *Drinking water standards governing drinking water quality and reporting requirements for public water systems* (30 TAC Chapter 290, Subchapter F: TCEQ RG-346). Retrieved July 19, 2010, from http://info.sos.state.tx.us/pls/pub/readtac$ext.ViewTAC?tac_view=5&ti=30&pt=1&ch=290&sch=F&rl=Y

Texas Commission on Environmental Quality [TCEQ]. (2010b). *Procedures to implement the Texas surface water quality standards* (RG-194. Water Quality Division). Retrieved July 22, 2010, from www.tceq.info/assets/public/legal/rules/rule_lib/adoptions/RG-194.pdf

Texas Commission on Environmental Quality [TCEQ]. (2010c). *Draft 2010 Texas 303(d) list*. Retrieved July 22, 2010, from www.tceq.state.tx.us/assets/public/compliance/monops/water/10twqi/2010_303d.pdf

Texas Commission on Environmental Quality [TCEQ]. (2010a). *Texas surface water quality standards* (Texas administrative code, title 30, part 1, chapter 307, rule 307.10). Retrieved July 22, 2010, from www.tceq.state.tx.us/assets/public/legal/rules/rule_lib/adoptions/07002307_ado_clean.pdf

Thermo Electron Corporation. (2006). *X-series ICP-MS operators course*. Cheshire, UK: Thermo Electron.

U.S. Environmental Protection Agency [U.S. EPA]. (1992). *Secondary drinking water regulations* (EPA 810/K-92-001). Retrieved February 15, 2010, from www.epa.gov/safewater/consumer/2ndstandards.html

U.S. Fish and Wildlife Service. (2003). *Proposed re-flooding and restoration of Bahia Grande, final draft environmental assessment*. Retrieved February 15, 2010, from www.fws.gov/southwest/refuges/texas/BahiaEA%20-%20Final%20Draft%20with%20Public%20Notice.pdf

U.S. Geological Society. (2002). *National map viewer*. Retrieved on June 8, 2007, from http://viewer.nationalmap.gov/viewer/

United Nations Environmental Programme [UNEP]. (2008). *Desalination: Resource and guidance manual for environmental impact assessments*. Manama, Bahrain/Cairo, Egypt: Regional Office for West Asia, World Health Organization, Regional Office for the Eastern Mediterranean.

Chapter 10
Evaluation of Hydraulic Residence Time Distribution (RTD) Characterization and Monitoring in a Constructed Channel Wetland in South Texas

Celia de la Mora Orozco and Kim D. Jones

Abstract Experiments comparing the use of an organic dye and sodium chloride as tracers were evaluated for their effectiveness in describing the hydrological conditions in a pilot-scale constructed channel wetland in semiarid South Texas, USA. This paper summarizes the significance, complexity, and challenges concerning the accuracy of the residence time distribution (RTD) determinations using organic and inorganic tracers in a pilot scale wetland evaluations. It was hypothesized that while an organic tracer had a much lower detection limit, its mass recovery due to sorption to organic surfaces might inhibit its performance and reliability. The pilot-scale RTD experiments were conducted using the organic Rhodamine WT (RWT) dye and sodium chloride (NaCl) as a tracers. Breakthrough curves were interpreted and differences in mass recovery suggested a slightly improved resolution of the wetland dispersive behavior when using RWT than when using NaCl as a tracer in these channel wetland applications. Mass recovery using RWT as a tracer was 67 % for the vegetated channel. Using NaCl, mass recovery was only 43 % in the vegetated channel. The calculated dispersion numbers were slightly higher using RWT (0.20 in the vegetated channel) than when using NaCl as a tracer (0.10 in the vegetated channel). Results from this research suggest some advantages of RWT as a tracer

C.d.l.M. Orozco
Institute for Sustainable Energy and the Environment, Texas A&M University-Kingsville, Kingsville, TX 78363-8202, USA

National Research Institute for Forestry, Agriculture and Animal Production, Mexico D.F., Mexico

Department of Environmental Engineering, Frank H. Dotterweich College of Engineering, Texas A&M University-Kingsville, Kingsville, TX 78363-8202, USA

K.D. Jones (✉)
Department of Environmental Engineering, Frank H. Dotterweich College of Engineering, Texas A&M University-Kingsville, Kingsville, TX 78363-8202, USA

Institute for Sustainable Energy and the Environment, Texas A&M University-Kingsville, Kingsville, TX 78363-8202, USA
e-mail: kim.jones@tamuk.edu

D. Ramirez et al. (eds.), *Environmental Sustainability Issues in the South Texas–Mexico Border Region*, DOI 10.1007/978-94-007-7122-2_10,

over NaCl due to the high mass recovery; however, both tracers gave similar quantitative results for wetland dispersion estimates.

Keywords Constructed wetlands • Dispersion number • Tracer testing • RTD estimation • Rhodamine dye • NaCl tracer testing

10.1 Introduction

The use of constructed wetlands for water quality improvement has been broadly studied (Borst, Riscassi, Estime, & Fassman, 2002; Jenkins & Greenway, 2005; Kadlec & Knight, 1996). Existing studies have mainly focused on pollutant removal from agricultural stormwater runoff and wastewaters. It is well known that numerous processes in wetland systems contribute to pollutant removal. Some of these processes that facilitate water quality improvement include biological (microbial uptake), physical (sedimentation), chemical (denitrification), and other less significant sinks (Coveney, Stites, Lowe, Battoe, & Conrow, 2002).

Kadlec and Knight (1996) described wetlands as a complex set of ecoreactors. A useful approach to evaluate the expected pollutant reduction in a wetland is to determine its retention time distribution (RTD) and apply a kinetic model. Chemical engineers often characterize RTD applications as an approximation as to how long individual molecules spend in the reactor (Fogler, 1992). The treatment process may be approximated by development of a material balance of component i as follows:

$$\begin{array}{l}\text{Net rate of}\\ \text{transport of } i\\ \text{throughout the}\\ \text{control volume}\end{array} + \begin{array}{l}\text{net rate of}\\ \text{transformation}\\ \text{of } i \text{ within the}\\ \text{control volume}\end{array} = \begin{array}{l}\text{net rate of}\\ \text{change of } i\\ \text{within the}\\ \text{control volume}\end{array} \tag{10.1}$$

From the mass balance above, an appropriate mathematical expression representing the specific reactor in question can be developed. Residence time experiments are essential when the exact flow pattern in a system is unknown (Werner & Kadlec, 2000).

In wetland systems, the treatment efficiency is governed to a great degree by the hydraulic residence time (HRT). Werner and Kadlec (2000) define the residence time distribution as a "distribution of the times that parcels of water spend in a constructed wetland" (p. 78). Theoretical hydraulic residence time (THRT) can be calculated using the water volume in the system divided by the volumetric flow rate. Thus, under ideal plug flow conditions the following equation can describe the hydraulic retention time.

$$\tau_n = \frac{V}{Q} \tag{10.2}$$

where:

V = volume of water in the wetland (m^3)
Q = volumetric flow rate (m^3/day)
τ_n = nominal retention time (days)

Under ideal plug flow conditions, the water moves in the form of a piston-like displacement. However, in a field-scale wetland system, the water flowing through it does not stay together as a single plug, but instead disperses and short circuits though the vegetation (Jenkins & Greenway, 2005). Under laminar conditions of flow, the velocity profile is uniform and the water moves through the system as if in parallel layers of flow. Persson (2000) states that uniform flow velocity can be achieved only in wetlands with a large length-width ratio, often described as channel wetlands. Water in wetlands systems does not move homogeneously as eddies are usually present (Persson). Thus, local velocity profiles provide limited information about the overall wetland hydraulic performance when compared to system wide tracer RTD experiments.

Besides the effect on pollutant removal efficiency in natural and constructed wetlands, hydrologic characteristics may affect vegetation composition, primary productivity, organic matter accumulation, and nutrient cycling (Mitsch & Gosselink, 2000). Deficient hydraulic designs may cause several wetland management and performance problems, such as low hydraulic efficiency and poor treatment (Gu, Chimney, Newman, & Nungesser, 2006). Parameters such as wetland shape, inlet and outlet hydraulic characteristics, bathymetry, type of vegetation, vegetation density and distribution in the system, and mixing can also influence wetland performance (Jenkins & Greenway, 2005). Also, it is well known that physical operational parameters such as wetland water depth and water flow rate can affect hydraulic performance and have a significant role in sediment deposition and nutrient removal (Brueske & Barrett, 1994). Borst et al. (2002) have suggested the importance of water depth for improving wetland efficiency, suggesting that the plant community is controlled by this variable. The sensitivity of wetland transport performance at different water depths and flow rates, using rhodamine WT (RWT) as a tracer, was evaluated by Holland et al. (2004). Water depth was controlled, ranging from 16.6 to 39.8 cm, and the flow rate limited to 1.2–3.2 Ls^{-1}. The RTD was found to be influenced by water depth changes, since as water depth increased short-circuiting became more evident and the degree of mixing also increased. Even though retention time distribution measurement is an approximation of how long water particles remain in a wetland system, RTD measurement has proven to be a useful tool to estimate wetland behavior and potential pollutant treatment (Konyha, Shaw, & Weiler, 1995; Worksdorfer, 2011). A longer HRT provides more time for sedimentation and chemical processes to occur, thus improving the treatment efficiency (Kjellin, Wörman, Johansson, & Lindahl, 2007). Typical recommended hydraulic retention times have varied from 4 to 15 days for treating wastewater in wetlands systems. However, Lin, Jing, and Lee (2003) carried out experiments in channel wetlands treating aquaculture effluent with residence times ranging from 1.1 to 8.4 days and reported significant nutrient (nitrogen and phosphorous) removal efficiencies with a residence time as short as 0.76 days. Their results showed that even at low residence times, removal efficiencies can be maintained at high levels. Significant biochemical oxygen demand (BOD) concentration reduction with increasing hydraulic retention time was reported by Tao, Hall, and Duff (2006) in a study designed to evaluate the effects of retention time and mass loading in a constructed wetland treating wood waste leachate. In this case, dissolved oxygen and redox potential decreased with increasing water depth. Temperature and influent BOD loading were important variables in treatment performance. The hydrologic condition is probably the most important factor for wetland performance success; however, it is rarely evaluated thoroughly and is difficult to measure (Mitsch & Gosselink, 2000). Persson (2000) has suggested that short-circuiting is the primary problem that decreases wetland efficiency. Short-circuiting can be defined as the wetland zones where water passes without treatment.

Jenkins and Greenway (2005) described how processes that take place in wetlands, such as filtration and sedimentation, are controlled by hydraulic conditions and vegetation relationships. Simulation models demonstrated the importance of fringing vegetation in increasing the retention time. Wetland size and shape were also important; however, the most significant effect in wetland efficiency was determined to be the vegetation distribution and density. Still, channeling may occur in wetlands due to occasional pathways formed by the vegetation spatial distribution (Kjellin et al., 2007).

10.2 Objective and Hypothesis

10.2.1 Goals and Objectives

In this investigation, the study objective was to evaluate the effectiveness, complexity, and challenges of making effective RTD measurements for treatment constructed wetlands in a semi-arid climate. Hydraulic conditions were characterized through RTD experiments using organic dye (RWT) and sodium chloride (NaCl) as tracers in pilot-scale wetlands. Three water depths were considered in these experiments (3, 6, and 9 in.). It was hypothesized that while RWT had a much lower detection limit, incomplete mass recovery due to sorption to organic surfaces could inhibit its performance and effectiveness. Therefore, an inorganic tracer might provide more accurate RTD measurements for engineering designs and comparisons.

10.2.2 Challenges for Retention Time Distribution Measurements

Tracer materials can be classified as either conservative or non-conservative. The principal advantage of using conservative substances is improved reliability as tracer losses in concentration between influent and effluent are minimized; the concentration may change only when new inflows are introduced in the water system. Sorption to organic matter is often cited as a tracer loss mechanism which can impact results (Lin, Debroux, Cunningham, & Reinhard, 2003). In the case of non-conservative substances, a great disadvantage may be substance decay or complexation. Decay is principally associated with chemical reaction, bacterial degradation, precipitation, and particle sedimentation out of the water column. The decay of the substance may be assumed to approximate a first-order reaction in which the rate of decay is also proportional to the mass at a given point in time. Thus, the non-conservative substance will decay and the downstream substance will drop exponentially as it approaches zero (Thomann & Mueller, 1987).

One of the most widely used techniques for evaluating the hydraulic characteristics in natural and constructed wetlands systems is tracer testing using fluorescent substances. The behavior of these dyes is often cited as a confounding factor in tracer test evaluations. Di Fazio and Vurro (1994) cited RWT as a fluorescent dye commonly utilized in tracer test evaluations in natural and artificial water systems. Usually, the main objective of the tracer test is to describe the physical mechanisms and hydrodispersive parameters in natural water systems. However, the authors point out that dye sorption is an issue to be considered when using RWT as a tracer in aquifers.

There are several tracers available for use derived from salts such as potassium bromide (KBr), potassium chloride (KCl), some radioactive compounds (primarily tritium), and most recently, pieces of DNA chains (Schmid, Hengl, & Stephan, 2004; Wood & Dykes, 2002). A major advantage of using RWT is its acceptance by environmental regulatory agencies due to its low environmental risk. In addition, it can be detected using a fluorescence detector at very low concentrations (around 0.01 μg/L). RWT is a large aromatic molecule with a fluorescence emission wavelength of 580 nm (Lin, Jing, & Lee, 2003). This material has a high fluorescence that is not commonly found in natural waters and is relatively stable in surface and ground water environments (Stern, Khanbilvardi, Alair, & Richardson, 2001). However, certain disadvantages of using RWT as a tracer have been pointed out, such as interferences when some natural fluorescence is present in the water and pH values below 5 that can reduce the dye fluorescence.

RWT sorption by plants, photochemical decay, and sediment sorption have been cited as major contributors to non-ideal RWT breakthrough behavior (Lin, Jing, & Lee, 2003). Some studies have reported low RWT mass recovery such as Lin, Debroux, Cunningham and Reinhard (2003) that in a field full-scale RTD experiment obtained only 29 % mass recovery. The contact time between water and sediment was suggested as a primary loss factor for the remaining 71 % RWT. Staff (2004) evaluated the nutrient removal efficiency in a pilot-scale constructed wetland using RWT as a tracer. In this case, wetlands were divided into two vegetated (A and B) and two unvegetated channels (A and B). The mass recovery for the two vegetated wetlands was 48 and 46 %, respectively, but for the two unvegetated wetlands was 49 and 63 %, respectively. Staff also suggested that the sorption and the shallow water level utilized in the experiment as a major factor contributing to dye loss. However, dispersive characterization was still possible. There are other studies that report higher mass recovery of RWT. A 72 % mass recovery was reported by Dierberg et al. (2005) when wetland hydraulic performance was evaluated using RWT. The loss of RWT was attributed to inflow upsets and the dead zones in the wetland. They also mentioned that the dye was probably lost by photolysis and adsorption. Another study evaluating sensitivity of wetland performance at different water depths and flow rates using Rhodamine WT as a tracer was used in a test carried out by Holland et al. (2004), who also reported a higher mass recovery ranging from 75 to 95 %, with an average of 84 %. These results suggest the suitability of RWT as a tracer in a wetland system under certain conditions.

Even though dye loss by degradation and sorption has been cited as an issue in RTD measurement, Stern et al. (2001) conducted field RTD experiments using RWT which were carried out in the Chillán River in Chile. The authors suggested that the experimental success may have been due to the short duration, and they reported no significant adsorption of RWT in field RTD experiments, although mass recovery was not mentioned. In their judgment, the results confirmed the use of dye as a suitable material for measuring RTD in natural wetlands. Although 100 % recovery of the dye is never expected, in the opinion of Stern et al. (2001) the use of RWT for tracer experiments still represents one of the most reliable methods of characterizing wetland hydrology. Even though RTD measurement using RWT can be an effective method, another possible disadvantage is that it is relatively expensive and in some cases the equipment needed for fluorescent measurement may not be available (Moore, 2003/2004).

Other tracers such as sodium chloride and radioactive tracers have also been suggested. Salt tracers such as bromide and sodium chloride are sometimes used in constructed wetlands tracer test experiments because they are inexpensive, inert, and do not present a risk to the environment. Weikel, Clinton, and Cathcart (2005) referred to the salt tracer method as the "slug test" salt conductivity tracer method, which consists basically of pouring a NaCl solution (slug) into the water system and the peak conductivity is measured downstream.

Although NaCl has been mentioned by some authors as a tracer that can be widely used in constructed wetlands, it is important to note that the literature does not support this. The majority of tracer tests using NaCl were found in ground water experiments and in measuring discharges in small streams. In fact, many NaCl studies were from experiments made in a laboratory setting and not in a field scale constructed wetland with plant material and outdoor seasonal conditions (Schmid et al., 2004). Although the salt tracer method has the advantage of being a cost-effective solution and is recognized as a benign method, it also has some disadvantages as reported by Schmid et al. In particular, stratification may have a significant effect in RTD experiments using NaCl as a tracer due to salt density variations.

Using high concentrations of inorganic salt as a tracer in surface waters may have other unforeseen consequences or concerns. High salinity may act as a stressor affecting wetlands plant dynamics and distribution in coastal areas where sea water level changes are typical. The growth of oligohaline marshes may be affected by salinity, yet many plants have the ability to recover after short periods of salinity exposure (Howard & Mendelssohn, 2000). Salinity might also affect the nutrient removal efficiency in constructed wetlands. Experiments carried out by Wu, Tam, and Wong (2008) report the negative impact on denitrification potential with increasing salinity from 15 to 30 ppt. The authors suggested that the salt enrichment was a major factor of microbial activity inhibition in soil. It has also been mentioned that salinity may inhibit the wetlands plants' growth and nitrogen and phosphorous absorption. However, this study also reported that *A. corniculatum* growth and uptake increased when salinity increased from 0 to 15 ppt suggesting this salinity as the optimum concentration.

In addition to plant growth and distribution effects, salinity may affect biota in the system. The salt injection effect on the drift of benthic macroinvertebrates was studied by Wood and Dykes (2002) in two streams under different flow conditions. The study results showed that some invertebrate taxa showed sensitivity to short periods of salt injection. However, the authors suggested that the invertebrate taxa sensitivity was a small disturbance with no significant temporal and spatial impact or plant stress. The authors also noted that the salt concentration peaks were low (under 100 mg/L) after mixing and suggested that biota were not exposed to toxic salt concentrations considering such a short time of exposure.

10.3 Experimental Design and Methods

A pilot-scale constructed wetland unit located at the United States Department of Agriculture's Kika de la Garza Plant Materials Center at the Texas A&M University – Kingsville field station. The unit consisted of two outdoor free water surface (FWS) constructed wetlands (W = 1.52 m, L = 4.27 m) that were divided longitudinally into two replicate cells (0.76 m × 4.27 m) using wood planks. The wetlands contained an approximate 30 cm sand layer with a double plastic layer used under the sand. Two cells contained two types of vegetation, *Avicenia germinans* (Black Mangrove) and *Scirpus americanus* (Olney Bulrush). Figure 10.1 shows a photo of both types of vegetation.

Fig. 10.1 Close up of TAMUK pilot-scale wetland channel showing the combination of bulrush and mangrove vegetation

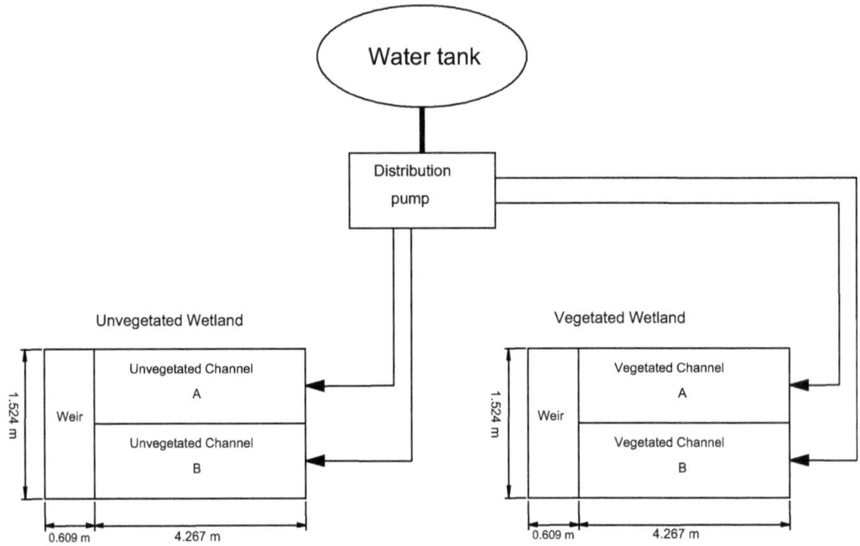

Fig. 10.2 Schematic of pilot scale wetland treatment system

The remaining two cells were non-vegetated. The inlet water was supplied by gravity from an influent tank that was controlled by a gate valve. A water depth of 7.62 cm and a retention time around 4 days were used to control the wetland hydraulic conditions. RTD experiments were conducted using RWT (20 % by weight) and sodium chloride as tracers. Figure 10.2 is a schematic of the wetland system.

10.3.1 Calculation of Rhodamine WT Solution Required

,An important factor in the RWT calculation is the desired peak concentration considering the highest and lowest concentrations of the dye used for the calibration curve. The dye concentration used to prepare the calibration curve has to be large enough for detection after recovery. The standards must be lower than this concentration (ASTM, 2000). The standards were prepared and measured with a Shimadzu RF-5301 PC spectrofluorophotometer. To estimate the amount of RWT (20 % by weight) solution needed in the experiment the following calculation was used (Eq. 10.3):

$$V_{LD} = \frac{W_{DD}}{W_{PC} \bullet D} \tag{10.3}$$

where

V_{LD} = volume of concentrated liquid dye required (mL)
W_{DD} = weight of dry dye desired (g)

W_{PC} = weight percent of dry dye in liquid dye concentrate, as specified by manufacturer
D = density of the dye concentrate (1.19 g/mL) for Rhodamine WT 20 % solution.

After the initial standard solution was prepared, the desired dilutions for calibration and field injection were easily prepared using Eq. (10.4):

$$V_1C_1 = V_2C_2 \tag{10.4}$$

where

V_1 = volume of desired tracer solution C_1, L
C_1 = desired concentration, mg/L of tracer C_1
V_2 = volume of standard tracer solution C_2, L
C_2 = concentration of standard tracer solution C_2, mg/L

10.3.2 Calculation of NaCl Required

The potential increment in electrical conductivity above the background concentration is a key factor to consider when using NaCl as a tracer, as is the sensitivity of the conductivity meter. The instrument detects the amount of salt in the solution during the wave passage. Changes in the peak conductivity can be used to characterize the mixing characteristics of the water system. At higher salt concentrations the peak wave will be higher which can mean less longitudinal dispersion; as a result, the injected salt solution concentration required can be lower (Moore, 2005). In this study, the NaCl concentration selected was sufficient to increase the conductivity to four times the background based on a pilot test with the influent water. However, using a lower concentration may not provide a representative RTD curve shape due to low signal-to-noise ratios in some cases. Similar calculations for tracer mass and volumes were made for the NaCl tracer solutions.

10.3.3 Tracer Solution Preparation

The RWT dye, 20 % by weight, was used in the experiments, and the specific gravity of the solution was 1.19. Equation 10.3 was used to calculate the dye required in the experiment. The volume used depended on the desired peak concentration. For these experiments 375 μL of the Rhodamine WT solution was diluted with 250 mL of water in a glass flask and introduced to the channel wetlands as pulses.

The sodium chloride tracer solution was prepared with respect to the background salinity concentration. The salinity was increased to four times the background concentration, as previous experiments demonstrated that lower amounts of salt may result in a non-typical curve shape. For this experiment 992 g of salt were diluted in 4 L of water. The solution was introduced as a pulse in each channel wetland. The flow rate for each channel wetland was maintained at 0.045 L/min keeping the hydraulic retention time around 4 days.

10.3.4 Experimental Design

Since a major objective of the study was to investigate the efficiency of nutrient removal at different wetland depths, the three depths were selected as an offshoot of the commonly mentioned optimal depth for constructed wetlands performance, <12 in. by Kadlec and Knight (1996), which was broken down into testing for 3, 6 and 9 in. of operating water depth. The nutrient levels for removal testing in subsequent experiments would be selected base on typical irrigation return flow levels experienced by operators in the region. A value of approximately 4 days of hydraulic retention time (HRT) was also selected for flow based on estimation of scale up of earlier studies of channel wetlands by Staff (2004). The tracer solution was added directly to the 500 gal storage tank which supplied a consistent influent flow through a metering valve attached to a perforated pipe header which spread out the flow at the front of the wetland channels.

Water samples for RWT experiments were collected in a 20-mL glass container from the end of each channel wetland every 3 h for day 1 and every 6 h thereafter. Samples were stored in the dark before analysis to avoid light decay. Sampling continued until the dye concentration was close to the background concentration or reached about 0.2 μg/L (2 % of the peak concentration) (ASTM, 2000). A Shimadzu RF-5301 PC spectrofluorophotometer was used to analyze the samples using an excitation wavelength of 486 nm and an emission of 558 nm. This instrument detects the luminescence of a substance at a specific wavelength since fluorescence varies linearly with dye concentration. In the case of the salt tracer experiment, a Hydrolab YSI Model 85 was installed at the end of each channel wetland and previously calibrated to measure the electrical conductivity and salinity every 30 min.

10.3.5 Estimation of Dispersion Coefficient and RTD

RTDs for pulse injection are typically a bell shaped curve where the effluent tracer concentration is plotted versus time. For this type of injection, the amount of tracer in the effluent is known as the concentration curve (*C* curve). The *C* curve provides idea characterization of how the dye injected distributes in the reactor with the time (Weber & DiGiano, 1996).

An estimate of total mass recovery can be calculated using Eq. 10.5.

$$M_T = Q \int_0^\infty C_{out}(t)dt \tag{10.5}$$

where

M_T = total mass recovery (mg)
Q = flow rate (m^3/s)
C_{OUT} = effluent concentration (mg/m^3)
t = time (s)

The C curve can be normalized into a dimensionless E function, $E(t)$, by dividing the effluent concentrations, C_{OUT}, by the hypothetical concentration $C_\Delta = M_T/V_R$, where V_R = reactor volume (m^3), and A_C is the total area under the normalized curve (Eq. 10.7)

$$E(t) = \frac{\frac{C_{out}(t)}{C_\Delta}}{\int_0^\infty \frac{C_{out}(t)}{C_\Delta}dt} = \frac{\frac{C_{out}(t)}{C_\Delta}}{A_C} \tag{10.6}$$

Since C_Δ in Eq. 10.6 is a constant, the area (A_C) under the normalized C_{out}/C_Δ curve can be calculated as

$$A_C = \frac{1}{C_\Delta} \int_0^\infty C_{out}(t)dt \tag{10.7}$$

where A_C has units of seconds.

The total area under the E curve is calculated as

$$\int_0^\infty E(t)dt = \frac{\int_0^\infty C_{out}(t)dt}{\int_0^\infty C_{out}(t)dt} = 1 \tag{10.8}$$

Thus, the E curve yields a time-normalized residence distribution such that

$$Fraction\ of\ elements\ with\ age < t = \int_0^\infty E(t)dt = 1 \tag{10.9}$$

The definition of hydraulic residence time (τ_n), known as detention time, is the ratio of the volume and flow ($\tau_n = V/Q$). Residence time distribution (RTD) can be defined as the distribution of times that particles spend in the reactor (wetland). The mean tracer residence time, τ_m, defined as the centroid of the RTD, is the average time that a tracer particle spends in the system.

The mean residence time, t_m, of a substance dissolved in a fluid is obtained by calculating the first moment of the centroid of the RTD curve, $E(t)$, about the origin.

$$t_m = \frac{\int_0^\infty tE(t)dt}{\int_0^\infty E(t)dt} \quad (10.10)$$

However, from Eq. 10.10, the denominator equals 1, therefore

$$t_m = \int_0^\infty tE(t)dt \quad (10.11)$$

The second moment of the centroid of the RTD, $\sigma_\tau{}^2$, about the mean gives the variance of the distribution. The variance quantifies the extent of spreading of the tracer "cloud" produced by a pulse input.

$$\sigma_t^2 = \int_0^\infty t^2E(t)dt - t_m^2 \quad (10.12)$$

A parameter, s^3, involving the third moment about the mean represents the skewness of the RTD. If the RTD is skewed to the left of the mean residence time, it suggests channeling in the reactor.

$$s^3 = \frac{1}{\sigma_t^{1.5}} \int_0^\infty (t - t_m)^2 E(t)dt \quad (10.13)$$

Using the previous calculations, the Péclet number (P_e), which is the ratio of the rate of transport by advection to the rate of transport by diffusion or dispersion, can be computed as follows:

$$\frac{\sigma^2}{t_m^2} = \frac{2}{P_e} - \frac{2}{(P_e^2)(1 - \exp^{P_e})} \quad (10.14)$$

Once the Péclet number is obtained, the effective dispersion coefficient, d, can be estimated with Eq. 10.15.

$$P_e = \frac{uL}{d} \quad (10.15)$$

where

P_e = Péclet number, ratio of rate of transport by advection to the rate of transport by diffusion or dispersion.
u = velocity (m/s)
L = length (m)
d = effective dispersion coefficient (m^2/s)

Since dimensional analysis is a useful approach to evaluate the character of significant parameters without respect to the units used in their derivation, the dimensionless form of this parameter is referred to as the dispersion number, D, as follows:

$$P_e = \frac{1}{D} \tag{10.16}$$

Dispersion in the system can result from various physical phenomena. The dispersion number is a descriptive term for characterizing flow and measuring the degree of mixing in the system. This number is important from an engineering design perspective because it helps to describe the nature of the flow in the system. When the dispersion number is large, dispersive transport dominates over advective transport in the system. Additionally, incorporating dispersion in a modeling application can delineate the system deviation from ideal plug flow. In theory, when the dispersion number increases the removal efficiency of certain pollutants may decrease with the additional transport and reduced retention time.

The spread of the distribution in the concentration curve depends on the variance, which is also known as the second moment about the mean. At greater variance values, there will be a larger spread in the distribution. The variance is also an important indicator of mixing. A large degree of mixing was observed in these studies using both tracers. The variance numbers obtained in these experiments were very similar using both tracer solutions.

10.4 Results and Discussion

The tracer experiments were conducted during September through October 2008 with water levels maintained daily at approximately 7.6 cm of depth and flow rate of approximately 0.045 L/min based on pump operating speeds calibrated through volumetric output testing. The experiments were carried out in duplicate for each tracer solution. The average results are shown in Fig. 10.3 with an average standard deviation of 3 %.

The tracer response curves obtained from using each of the tracers followed a characteristic pattern for similar wetland RTD experiments (Werner & Kadlec, 2000). Figure 10.3 shows the concentration-versus-time curves for the RWT experiments. The peak concentration was slightly lower in the vegetated channel

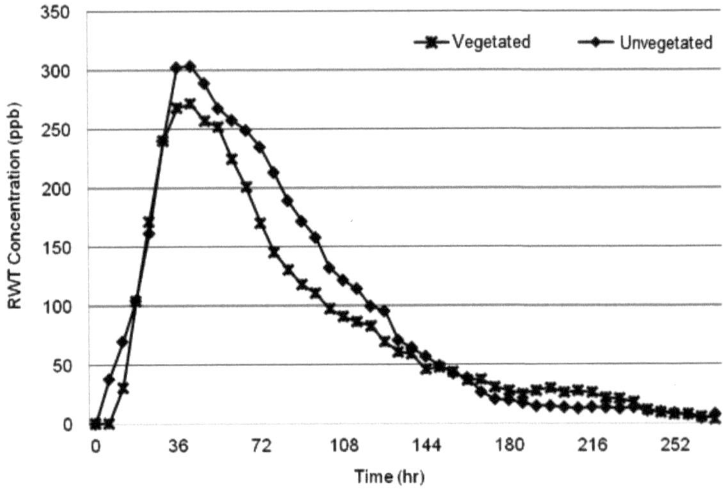

Fig. 10.3 The average effluent concentration curve from RTD experiment carried out on the vegetated and control constructed channel wetlands using Rhodamine WT as a tracer

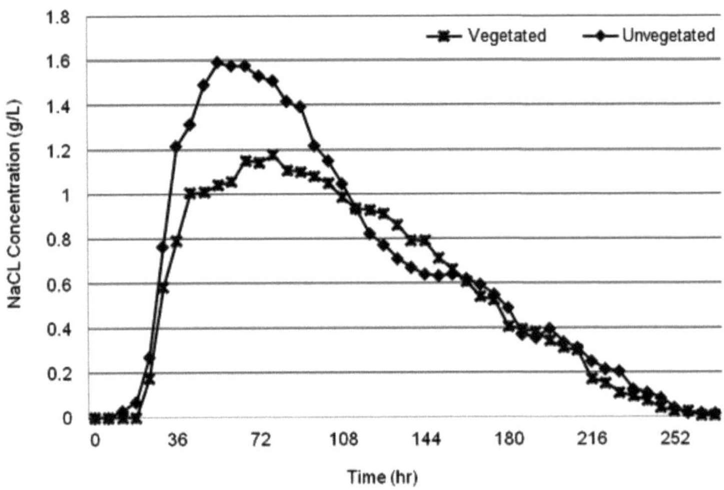

Fig. 10.4 The average effluent concentration curves from RTD experiments carried out on the vegetated and control constructed channel wetlands using sodium chloride as a tracer

than in the control channel, which suggested some potential RWT sorption to organic matter.

Using NaCl as a tracer (Fig. 10.4) produced an evident difference between the peak concentration in the control and vegetated channels. The NaCl peak concentration in the vegetated channel was slightly lower than the peak in the control channel. One complexity in interpreting these results is that high salt concentrations tend to stratify. This behavior was observed in laboratory experiments carried out

Table 10.1 Calculation results for average wetland RTD measurements and the standard deviation for the duplicate

Channel	Mass recovery (%)		Mean residence time = t_m (h)		Variance = σ^2 (h^2)		Dispersion number		Skewness = s^3 (h)	
wetland	RWT	NaCl	RWT	NaCl	RWT	NaCl	RWT	NaCl	RWT	NaCl
Control	**74.280**	**47.950**	**75.657**	**100.984**	**33.764**	**37.862**	**0.205**	**0.108**	**7.100**	**6.283**
STDVE	2.571	0.495	0.793	1.533	1.665	2.966	0.011	0.004	0.079	0.233
Vegetated	**64.196**	**40.920**	**79.953**	**105.850**	**38.270**	**36.089**	**0.133**	**0.052**	**7.305**	**5.713**
STDVE	4.595	0.311	5.679	7.142	0.468	2.717	0.018	0.004	0.373	0.877

by Schmid et al. (2004). A further consideration is the low flow rate impact on the salt tracer. The slow water movement may enhance the salt stratification, resulting in more complex transport behavior. Plants may further decrease the flow velocity and strongly influence the tracer response.

Table 10.1 summarizes the average results for mass recovery, mean residence time, and variance obtained from the experiment at 7.6 cm of water depth.

10.4.1 Mean Residence Time Distribution

Using sodium chloride solution as a tracer, the mean residence times in the control and vegetated channels were calculated to be 4.20 and 4.40 days, respectively. These values were 0.20 and 0.43 days higher, respectively, than the estimated theoretical hydraulic residence (V/Q) time of about 4 days. Using the RWT solution, the mean residence times in the control and vegetated channel wetlands were calculated to be 3.17 and 3.34 days, respectively, which were 0.63 and 0.46 days less, respectively, than the hydraulic theoretical residence time. Thus, the experimental tracer residence time for the RWT tracer solution was slightly shorter than the nominal residence time and it was slightly longer using NaCl. Since adsorption or stratification of the tracer may occur together with short-circuiting, the experimental tracer mean residence time may turn out to be longer or shorter than the hydraulic residence time. Persson and Wittgren (2003) noted that the mean tracer residence time, τ_m, is typically less than the theoretical hydraulic residence time, τ_n. Additionally, the experimental residence times were shorter in the control channel than in the vegetated channel using either the NaCl or the RWT tracer. These results using RWT as tracer were similar to those of Staff (2004). The presence of vegetation in wetlands may affect the flow resulting in increasing the wetland hydraulic efficiency. Visual observations and measurements estimated that vegetation covered approximately 60 % of the areal space in the vegetated wetland at the time of this experiment. For the results using NaCl as the tracer, the calculated average residence time appeared to be slightly longer for both of the vegetated and control channels. The tendency for NaCl to stratify mentioned by Schmid et al. (2004) and the depths used in this experiment may have potentially caused some of these variations with respect to the RWT results. However, when irregular patterns of vegetation are present some zones may be created where mixing can be diminished and the hydraulic efficiency may decrease (Holland et al., 2004).

10.4.2 Mass Recovery Calculations

A mass balance of RWT was performed comparing the tracer mass added and the tracer mass recovered from the effluent. Mass recovery of 100 % would not be expected when using the RWT tracer due to its affinity for organic matter. A comparison demonstrated 74 and 64 % of mass recoveries of RWT in the control and vegetated channels, respectively. Similar results were obtained by Dierberg et al. (2005), who reported a 72 % mass recovery where RWT was used as a tracer in constructed wetlands. Holland et al. (2004) reported a slightly higher average mass recovery of approximately 84 %.

Mass recoveries using NaCl as a tracer were approximately 30 % less than when using RWT, or 48 and 41 % in the control and vegetated channels, which could potentially be explained due to density driven stratification of the tracer. Even though degradation and sorption have been reported as the main causes of RWT loss in RTD experiments, the current findings suggested that the RWT may be a more advantageous tracer than the NaCl solution under certain short duration experiments from a mass balance perspective in such wetlands. Konyha et al. (1995) found similar results as the one obtained in this study.

10.4.3 Wetlands Dispersion Results Estimation

The dispersion numbers using sodium chloride and RWT tracer were 0.10 and 0.20 in the control channel, respectively. The vegetated channel dispersion numbers were 0.05 and 0.13 using sodium chloride and RWT, respectively. Larger wetland dispersion numbers were reported by Dierberg et al. (2005); however, their experiments were run at varying conditions of detention time and flow rate. On the other hand, Persson and Wittgren (2003) noted that an average wetland dispersion number is typically between 0.1 and 0.4, and the results reported here are within this range except for the vegetated channel measurement with NaCl. Moreover, the curve shapes for the experiments in these channels closely approximate the Gaussian bell-shaped curve. Higher dispersion creates a large variance in the distribution curve (concentration versus time), which also suggests the presence of some parcels of water moving quickly through the channel and some traveling more slowly. A low degree of mixing is often desired to decrease short-circuiting and increase wetland removal performance. A lower dispersion number in the vegetated wetland suggests that vegetation may serve as a protective barrier from environmental mixing conditions caused by wind and temperature driven changes.

Thus, in order to diminish dispersion in channel wetlands the presence of vegetation may be highly desirable improving the system efficiency. The control channels had larger dispersion numbers that confirmed the presence of plants as a desirable constituent in the channels to reduce some longitudinal and transverse mixing.

10.4.4 Wetland Variance

Variance values for the control and vegetated channels were 38 and 36 h, respectively, using NaCl as a tracer. Using RWT the values obtained were 34 and 38 h for the control and vegetated channels, respectively. Staff (2004) cited wind driven mixing as a potential cause of the higher variance in open channels. In this particular case, variance numbers were very similar using both tracer solutions; however, Staff also suggested that the presence of vegetation may be an important element protecting the system against enhanced mixing due to wind and solar heating.

10.4.5 Skewness

The magnitude of the skewness value, s^3, represents the extent that a distribution is skewed in any direction in reference to its mean (Weber & DiGiano, 1996). When the s^3 value is positive, it means the RTD is skewed toward times greater than the mean value. Skewness toward times shorter than the mean residence time might be an indication of channeling in the system. The skewness results were less than the mean which suggested the possibility of channeling in the system.

10.5 Conclusions

RTD estimation for constructed wetlands is a challenging task. Some of the potential issues include incomplete tracer mass recovery and tracer detection limits. Caution is necessary when choosing the tracer concentration due to detection limit concerns and losses. This work presents some new data that demonstrates the utility of both organic (RWT) and inorganic (NaCl) tracers for wetland hydraulic characterization and has provided some specifics for their use and interpretation for semiarid zone applications.

Mass recovery is of particular interest when measuring RTD due to its potential influence on accuracy. In this study, organic RWT material demonstrated better mass recovery than the inorganic NaCl solution. This result suggests that RWT may be a more reliable tracer for wetlands, even though it may sorb to organic matter. Due to the potential density stratification of sodium chloride, it may be a preferred tracer in systems where flow rates are high, such as rivers, since physical forces are stronger than in wetlands where water usually flows at low flow rates. Hydraulic condition evaluations in a pilot-scale wetland may contribute essential information to wetlands design by assessing how RTD, mass loading rate, and water depth affect nutrient removal efficiency. Constructed channel wetlands may have potential for the mitigation of nonpoint source pollution particularly in South Texas and Northern Mexico. This study highlights methods that engineers and scientists could use to determine residence time distributions associated with various wetland designs in these semiarid climates. Future research plans will confirm the findings shown here by measuring dispersion at different water depths in the pilot-scale constructed wetlands.

Acknowledgements This material is based upon work supported by the Center of Research Excellence in Science and Technology – Research on Environmental Sustainability of Semi-Arid Coastal Areas (CREST-RESSACA) at Texas A&M University– Kingsville (TAMUK) through a Cooperative Agreement (No. HRD-0734850) from the National Science Foundation (NSF), John Lloyd-Reilley of the USDA Kika de la Garza Plant Materials Center, and the South Texas Environmental Institute at Texas A&M University-Kingsville. Any opinions, findings, and conclusions or recommendation expressed in this materials are those of the authors and do not necessarily reflect the views of the National Science Foundation, John Lloyd-Reilley of the USDA Kika de la Garza Plant Materials Center, and/or the South Texas Environmental Institute at Texas A&M University – Kingsville.

References

ASTM. (2000). Standard test method for open-channel measurement of time of travel using dye tracers. In *Annual book of ASTM standards*. Philadelphia, PA: ASTM.

Borst, M., Riscassi, A., Estime, L., & Fassman, E. (2002). Free-water depth as a management tool for constructed wetlands. *Journal of Aquatic Plant Management, 40*, 43–45.

Brueske, C., & Barrett, G. (1994). Effects of vegetation and hydrologic load on sedimentation patterns in experimental wetland ecosystems. *Ecological Engineering, 3*(4), 429–447.

Coveney, M., Stites, D., Lowe, E., Battoe, L., & Conrow, R. (2002). Nutrient removal from eutrophic lake water by wetland filtration. *Ecological Engineering, 19*(2), 141–159.

Di Fazio, A., & Vurro, M. (1994). Experimental tests using rhodamine WT as tracer. *Advanced Water Resources, 17*(6), 375–378.

Dierberg, F., John, J., Juston, J., DeBuska, A., Pietro, K., & Gub, B. (2005). Relationship between hydraulic efficiency and phosphorus removal in a submerged aquatic vegetation dominated treatment wetland. *Ecological Engineering, 25*(1), 9–23.

Fogler, H. (1992). *Elements of chemical reaction engineering*. Upper Saddle River, NJ: Prentice Hall.

Gu, B., Chimney, M., Newman, J., & Nungesser, M. (2006). Limnological characteristics of a subtropical constructed wetland in south Florida (USA). *Ecological Engineering, 27*(4), 345–360.

Holland, J., Martin, F., Granata, T., Bouchard, V., Quigley, M., & Brown, L. (2004). Effects of wetland depth and flow rate on residence time distribution characteristics. *Ecological Engineering, 23*(3), 189–203.

Howard, J., & Mendelssohn, A. (2000). Structure and composition of oligohaline marsh plant communities exposed to salinity pulses. *Aquatic Botany, 68*(2), 143–164.

Jenkins, G., & Greenway, M. (2005). The hydraulic efficiency of fringing versus banded vegetation in constructed wetlands. *Ecological Engineering, 25*(1), 61–72.
Kadlec, R., & Knight, R. (1996). *Treatment wetlands*. Boca Raton, FL: Lewis Publishers.
Kjellin, J., Wörman, A., Johansson, H., & Lindahl, A. (2007). Controlling factors for water residence time and flow patterns in Ekeby treatment wetland, Sweden. *Advanced Water Resources, 30*(4), 838–850.
Konyha, K., Shaw, D., & Weiler, K. (1995). Hydrologic design of a wetland: Advantages of continuous modeling. *Ecological Engineering, 4*(2), 99–116.
Lin, Y., Debroux, J., Cunningham, J., & Reinhard, M. (2003). Comparison of rhodamine WT and bromide in the determination of hydraulic characteristics of constructed wetlands. *Ecological Engineering, 20*(1), 75–88.
Lin, Y., Jing, S., & Lee, D. (2003). The potential use of constructed wetlands in a recirculating aquaculture system for shrimp culture. *Environmental Pollution, 123*(1), 107–113.
Mitsch, W., & Gosselink, J. (2000). *Wetlands*. New York: Wiley.
Moore, R. (2003/2004). Introduction to salt dilution gauging for streamflow measurement: Part 1. *Streamline Watershed Management Bulletin, 7*, 20–23.
Moore, R. (2005). Slug injection using salt in solution. *Streamline Watershed Management Bulletin, 8*, 1–6.
Persson, J. (2000). The hydraulic performance of ponds of various layouts. *Urban Water, 2*(3), 243–250.
Persson, J., & Wittgren, H. (2003). How hydrological and hydraulic conditions affect performance of ponds. *Ecological Engineering, 21*(4–5), 259–269.
Schmid, H., Hengl, A., & Stephan, U. (2004). Salt tracer experiments in constructed wetland ponds with emergent vegetation: Laboratory study on the formation of density layers and its influence on breakthrough curve analysis. *Water Research, 38*(8), 2095–2102.
Staff, M. (2004). *Modeling the performance of channel wetlands design to treat aquaculture effluents with plug flow reactor and plug flow with dispersion approach*. Master's thesis, Texas A&M University-Kingsville, Kingsville, TX.
Stern, D., Khanbilvardi, R., Alair, J., & Richardson, W. (2001). Description of flow through a natural wetland using dye tracer tests. *Ecological Engineering, 18*(2), 173–184.
Tao, W., Hall, J., & Duff, S. (2006). Performance evaluation and effects of hydraulic retention time and mass loading rate on treatment of woodwaste leachate in surface-flow constructed wetlands. *Ecological Engineering, 26*(3), 252–265.
Thomann, R., & Mueller, J. (1987). *Principles of surface water quality modeling and control*. New York: Harper Collins.
Weber, W., & DiGiano, F. (1996). *Process dynamics in environmental systems*. New York: Wiley.
Weikel, J., Clinton, J., & Cathcart, J. (2005). *Monitoring protocol for reporting impacts of the salt tracer methodology on stream biota*. Salem, OR: Forest Health and Monitoring Unit, Private and Community Forests Program, Oregon Department of Forestry. Retrieved July 25, 2008, from http://www.oregon.gov/ODF/privateforests/docs/SaltTracerProtocol.pdf?ga=t
Werner, T., & Kadlec, R. (2000). Wetland residence time distribution modeling. *Ecological Engineering, 15*(1–2), 77–90.
Wood, P., & Dykes, A. (2002). The use of salt dilution gauging techniques: Ecological considerations and insights. *Water Resources, 36*(12), 3054–3062.
Worksdorfer, C. (2011). Tracer test in a settling pond: The passive mine water treatment plant of the 1 B Mine Pool, Nova Scotia, Canada. *Mine Water Environment, 30*(2), 105–112. doi:10.1007/s10230-011-0147-3. Published online.
Wu, Y., Tam, N., & Wong, M. (2008). Effects of salinity on treatment of municipal wastewater by constructed mangrove wetland microcosms. *Marine Pollution Bulletin, 57*(6–12), 727–734.

Chapter 11
Recovery of Aquaculture Treatment Constructed Wetlands Function After Prolonged Flood Inundation Events in South Texas

Kim D. Jones and Brian Dyson

Abstract Constructed wetlands have been evaluated for many years as a means of low cost water treatment; however studies of performance evaluation and recovery after system upsets are not often reported. The Loma Alta Shrimp Aquaculture Facility (LASAF), located on the El Sauz Ranch in South Texas employs an innovative re-circulating aquaculture-wetland system where pond effluent is filtered through a constructed wetland and then re-circulated back into the shrimp production ponds. The goals of this study were to evaluate the relative water quality treatment ability for several wetland areas and evaluate the recovery of nutrient removal activity after major flooding damage and inundation in 2002. The practical implications of these results were incorporated into the design of a new constructed wetland that treats pond effluent in a sequential manner based on water depth. Despite variable loadings from aquaculture pond discharges, a steady state operation developed prior to harvest operations can be used to evaluate the nutrient removal performance of the wetlands, which can be determined through a change point statistical technique. The flooding events caused by storms and operator management requirements in 2002 resulted in diminished vegetation and wetland performance which was able to be restored in 2003, with the growth of primarily submerged vegetation. The smaller (8 ha) Phase III sequential basin design appeared to be able to effectively treat aquaculture effluent as well as the larger

K.D. Jones (✉)
Department of Environmental Engineering, Frank H. Dotterweich College of Engineering, Texas A&M University-Kingsville, Kingsville, TX 78363-8202, USA

Institute for Sustainable Energy and the Environment, Texas A&M University-Kingsville, Kingsville, TX 78363-8202, USA
e-mail: kim.jones@tamuk.edu

B. Dyson
Department of Environmental Engineering, Frank H. Dotterweich College of Engineering, Texas A&M University-Kingsville, Kingsville, TX 78363-8202, USA

U.S. Environmental Protection Agency, Washington, DC, USA

D. Ramirez et al. (eds.), *Environmental Sustainability Issues in the South Texas–Mexico Border Region*, DOI 10.1007/978-94-007-7122-2_11,

(19 ha) dual system Phase I and II wetlands. Close attention to water depth and recirculation rate management suggest a linkage to optimal performance for water treatment targets. It is suggested that techniques such as change-point analysis can help confirm as to when the retention time significantly changes in a dynamic wetland system as a result of changing water depth and provide insight as to the nature of its hydraulic efficiency.

Keywords Constructed wetlands • Aquaculture • Wetland restoration • Wetland recovery

11.1 Introduction

The total value of U.S. aquaculture is about $6 billion and the U.S. industry supports at least 181,000 full-time jobs. Aquaculture continues to be the fastest growing sector of U.S. agriculture (Treece, 2005). While the growth of the global aquaculture industry holds tremendous potential for alleviating food shortages in many areas of the world, the production of many fish species in ponds can create some water quality degradation including elevated levels of suspended solids, biochemical oxygen demand, and nutrients (Paez-Osuna, 2001). As a result, frequent water exchange becomes necessary, increasing the risk of environmental pollution and economic losses due to disease, prompting the need for more bio-secure production methods that reduce water exchange (Moss, 2002).

One viable method of improving water quality and reducing water exchange for aquaculture applications is the use of constructed wetlands. Costa-Pierce (1998) used an aquaculture-wetland system (AWS) to produce fish from treated municipal effluent. Lin, Jing, Lee, and Wang (2002) and Coveney, Stites, Lowe, Battoe, and Conrow (2002) investigated nutrient removals in constructed wetlands as treatment systems. Tilley, Badrinarayanan, Rosati, and Son (2002) reported nitrogen removals of 86–98 % for ammonium nitrogen and 95–98 % for total inorganic nitrogen for a re-circulating AWS. Aquaculture sludges treated with constructed wetlands have been reported to have significant removals of total suspended solids, chemical oxygen demand, total Kjeldahl nitrogen, and phosphorus (Summerfelt, Adler, Glenn, & Kretschmann, 1999). The development of this eco-technology has also demonstrated the need for more effective models to predict performance and enhance designs (Carleton, 2002; Kadlec, 2000; Kadlec & Knight, 1996).

Aquaculture is a growth industry for the U.S. Recent production has reached almost 900 million pounds annually for all species, but this has also resulted in stricter effluent guidelines to mitigate the impact of farm discharges (Jensen, 2005). The Loma Alta Shrimp Aquaculture Facility (LASAF) is located in the coastal region of semi-arid South Texas. In 2000, LASAF operated 8.1 ha of shrimp aquaculture ponds treated by 7.7 ha of constructed wetlands. Effluent from the ponds was passed through the wetland and re-circulated to the ponds. The benefit of this mode of operation is reduced intake of source water and no discharge of effluent to receiving waters. Tilley et al. (2002) originally studied the LASAF system in 2000 and reported reductions in total phosphorus, total suspended solids, and inorganic suspended solids on the order of 31, 65, and 76 % respectively. In the following year, production was increased to 36 ha of ponds treated by 18.8 ha of constructed wetlands. The wetland area was increased by the connection of an 11.2 ha wetland to the existing wetland by a channel causing the effluent to pass through the wetlands in series before being re-circulated to the ponds (Fig. 11.1). The most recent wetland (Phase II) has an average depth of 0.5 m and is dominated by widgeon grass (*Ruppia maritima*) a submerged aquatic plant. The older wetland (Phase I) has an average depth of 0.3 m and was originally dominated by cattails (*Typha* spp.). The flow of effluent is through the Phase II system, followed by Phase I and then back into the ponds. Figure 11.1 is a schematic diagram of the water movement for these two wetlands. In 2003, additional ponds were added and a wetland of a sequential design (shallow zones of emergent vegetation followed by deeper zones of floating vegetation) was constructed of approximately 11 ha in size. This wetland has been labeled Phase III. Figure 11.2 is a diagram of its water flow pattern. Supply water flows through the ten new ponds through the Phase III wetlands and then back to the supply canal for recirculation.

The original objectives of this study were to evaluate the removals of nutrients in aquaculture constructed wetlands in the semiarid climate of South Texas, compare the removal performance of three wetland designs, and examine the effect of water depth and vegetation type on wetland performance over four seasons of aquaculture production and harvesting. The three types of wetlands employed included field scale wetlands of a shallow depth (Phase I, 6–12″ water depth) with primarily emergent vegetation, a deeper constructed wetland (Phase II, 12–30″ water depth) with primarily submerged vegetation, and a constructed wetland built with alternating zones of shallow and deeper depths (Phase III) to allow sequential treatment by both emergent and submerged plant material and enhance nutrient mixing by flow through deeper basins. LASAF filters seventeen 2.8-ha and four 2-ha ponds with three wetlands: Phase I (7.6 ha), Phase II (11.2 ha) and Phase III (7.5 ha). Phase I and II wetlands are diagramed in Figs. 11.1 and 11.2 shows the layout of the Phase III wetlands.

Also, as heavy rains and operational necessity caused large volumes of water to be stored in the wetlands in 2002, the recovery of the removal efficiency for the wetlands systems was also evaluated after this event. During the 2002 season, most water levels typically kept at 6–8 in. (15–20 cm), were raised up to 2.5 ft (75 cm). This caused a significant reduction in both emergent and submerged vegetation, and reduced wetland efficiency in removing nutrients. A subsequent objective of this work became the evaluation of the recovery in performance of the constructed wetlands in removing nutrients for effective water recirculation and reuse after the major upset in 2002.

For most of the year, the wetland system operates under approximate steady state conditions during the shrimp growing season as a steady stream of water is trickled from several ponds through the wetlands and re-circulated back to the ponds. But once harvest begins the state of the wetland and its function can become more dynamic. The identification of this steady state regime is important to effectively interpret treatment efficiency from sampling measurements collected along the

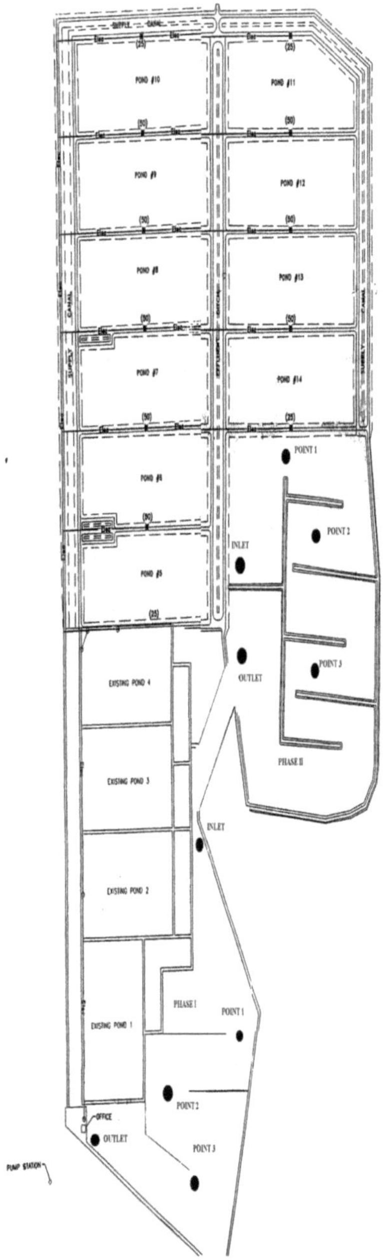

Fig. 11.1 Loma Alta Facility aquaculture-wetland system diagram showing production ponds connected to Phase I and Phase II wetlands

length of the large wetland systems. LASAF is permitted by the Texas Commission on Environmental Quality (TCEQ) and only discharges water from the system at the end of the season, meaning that prior to discharge effluent can only leave the system via evapo-transpiration, groundwater seepage or occasionally back-filling

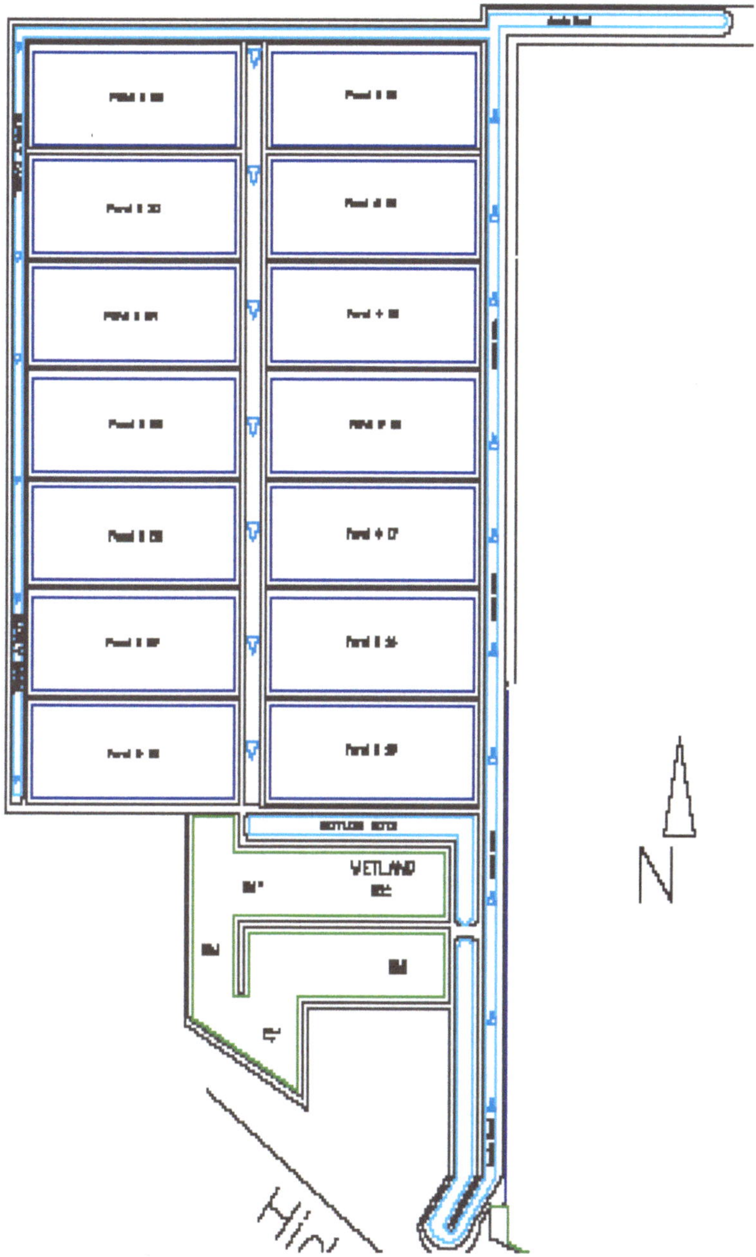

Fig. 11.2 2003 Loma Alta Facility aquaculture-wetland system showing Phase III expansion ponds and Phase III wetlands

an empty pond. Once pond releases begin in harvest season, it becomes difficult to use standard statistical tests to evaluate wetland performance due to uncertainties associated with proper sampling for such a dynamic system. Change-point analysis can be utilized, however, to determine the point at which the state of wetland operation was altered, and thus the time at which simple statistical techniques will become less applicable. Prior to the onset of significant harvesting and pond releases, the performance of the various wetlands throughout the growing seasons can be evaluated as they are operated in an approximate steady state mode.

11.2 Methodology and Analysis

11.2.1 Sample Collection and Analysis

Water samples were collected on a weekly basis from June to November during the shrimp production seasons for 2001 through 2004. Inlet and outlet grab samples were collected within each wetland far enough away from the inlet/outlet structures to avoid channeling effects. Statistical analysis of wetland function before release was performed using the Mann-Whitney U test for removals. For the whole season, overflow measurements were made at the wetland outlets and flow calculated using a sharp-crested weir formula. Depth was measured at the same location as the grab samples.

Each wetland had five sampling sites; inlet, outlet, and three locations within the wetland following the flow of water (Fig. 11.1). Water samples were collected at mid-depth using 1-l HDPE bottles between the hours of 10 am and 12 pm. Samples collected for nitrogen testing were preserved to a $pH < 2$ with H_2SO_4. All samples were placed on ice and transported to the laboratory. Preserved samples were then neutralized, filtered, and tested within 48 h.

Temperature, salinity, and dissolved oxygen were measured *in situ* using an YSI Model 85 handheld probe. An YSI Model 63 handheld probe was used to measure pH *in situ*. Laboratory analyses of ammonia (NH_3-N) and total phosphorus (mg P/l) were performed with the HACH DR2000 (HACH, 1991) using HACH Methods 10023 for ammonia (chosen due to effluent salinity) and 8190 for total phosphorus. Five-day biochemical oxygen demand (BOD_5), total, inorganic, and suspended solids were measured following Standard Methods (APHA, 1992) using Method 5210B for BOD_5 and Methods 2450D-E for suspended solids.

11.2.2 Statistical Analyses and Methods

Change Point Analysis to determine approximate dynamic operating condition was performed by analyzing the change in depth and flow over the whole season using Change Point Analyzer® (version 2.3) (Taylor, 2000). Flow was chosen because it directly affects residence time-a key determinant in treatment efficiency and depth because increasing water depth can decrease hydraulic efficiency (Holland et al., 2004). This analytical method combines cumulative sum charts (CUSUM) (Hinkley, 1971) with bootstrapping (Hinkley & Schechtman, 1987), is robust to outliers, and provides a confidence level for change points identified (Taylor).

One-way ANOVA and paired-mean difference testing were performed using randomization testing. This was done with the Resampling Stats (Resampling Stats, 2002) add-in Version 2.0 for Microsoft EXCEL®. Post hoc testing for ANOVA was performed using the Scheffe' test also by randomization. The advantages of using randomization are two-fold (Manly, 2001); (1) there is no need to assume a particular distribution of sample data as in parametric testing (n = 15 for current sample data); randomization creates a distribution of the statistic in question from which significance testing can be performed, and (2) the original data may be used instead of ranks as in non-parametric testing.

11.3 Results and Discussion

11.3.1 Flow Rates and Retention Times

Harvesting operations were begun at different times for each year based on market considerations and operational constraints. To summarize, significant harvest and pond discharges were initiated on August 8 in 2001, September 3 in 2002, July 3 in 2003, and August 4 in 2004. This was a good first estimate of when wetland system operation influent and effluent conditions began to be subjected to more dynamic loading. Market conditions and weather events mandated random harvesting in 2005 for the operator and it is doubtful a quasi-steady state period was experienced.

The period from first release to change-point is a transition period that can give insight into the hydraulic efficiency of these wetlands. The transition period can be viewed as a functional change in hydraulic residence time (HRT). Prior to the harvest periods, the flow rates through the systems were much slower than subsequent to pond harvesting allowing for more effective retention time and nutrient removal. For example, in 2003, the average full system flow rate prior to August was 0.41 MGD and subsequent to that time the flows were increased to an average of 3.82 MGD. In 2004, the average flow rate increased from 1.33 MGD before August to 3.36 MGD after August 1. The nominal retention times (V/Q ratio) were estimated at 23 days for Phase II, 16 days for Phase I and 17 days for Phase III.

Significantly, the wetlands vegetation type and distribution changed over the study period. Table 11.1 summarizes the wetland plant species dominance and estimated coverage based on visual observations. Figure 11.3 depicts the significant submerged wigeongrass vegetation typically found in the deeper zones of Phase II, and Fig. 11.4 shows the emergent Olney bulrushes in Phase III.

Table 11.1 Dominant vegetation and coverage estimates for each wetland

	Veg type/coverage		
Wetland season	Phase I	Phase II	Phase III
2001	Cattails/30 %	Widgeon grass/50 %	–
2002	Saltmarsh bulrush/5 %	Widgeon grass/5 %	–
2003	Saltmarsh bulrush/10 %	Widgeon grass/20 %	Olney bulrush/10 %
2004	Saltmarsh bulrush/5 %: widgeon grass/5 %	Widgeon grass/50 %	Olney bulrush/10 %

Fig. 11.3 Photo of submerged wigeon grass vegetation prevalent in the deeper wetland zones of Phase I and Phase II wetlands

Fig. 11.4 Photo of emergent Olney bulrush vegetation growing in the Phase III sequential wetland shallow zone cells

It is noteworthy that the dominant vegetative species in this region can fluctuate for a large constructed wetland apparently due to fluctuations in rainfall, nutrients, salinity and other environmental factors. Rigorous vegetation management for such large areas was cost prohibitive. The operator had a practice of seasonally drying

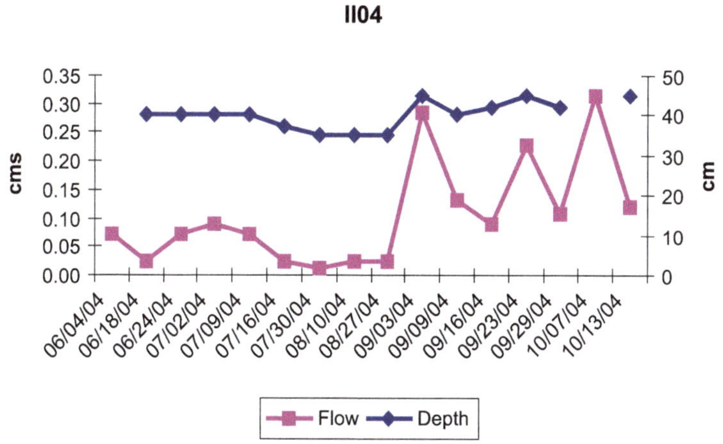

Fig. 11.5 Diagram of time versus wetland water circulation rate (cm/s) and water depth (cm) for the Phase II wetland in 2004 showing the effect of the onset of harvest conditions on wetland conditions on 9/3/04

out the ponds and the wetlands for 2 weeks each in attempts to control shrimp pathogens. The cattails became stressed and receded in 2001 apparently due to high water from heavy rainfall and salinity increases earlier in the season. The saltmarsh bulrush reestablished itself but it did not appear as robust as the cattails and the emergent plant biomass was considerably less than when cattails were dominant. The submerged widgeon grass always dominated the deeper Phase II section, and the Olney bulrush was planted along with some mangroves in the shallow sections of Phase III. The submerged widgeon grass slowly began to dominate even the older sections with emergent vegetation.

Change-point analysis was performed for the treatment wetlands based on flow and inlet depth and found to be reasonable and consistent with the original estimate in change from quasi-steady state to dynamic conditions. For example, Fig. 11.5 shows the effect of pond releases for the Phase II wetland over the 2004 operating season, by displaying the flow (cm/s) and water depth (cm) for the production and harvest periods. The apparent change to dynamic wetland operating conditions on 9/3/04 was also verified by this analysis.

11.3.2 Initial Phase I and II Wetlands Evaluation and New Sequential Design

In 2001, initial removal efficiency analysis showed the Phase I and II wetlands exhibiting different treatment abilities with respect to particular pollutants. Although the effluent eventually passes through both wetlands, the two wetlands are physically separated and were evaluated as functioning independently. In 2000,

Table 11.2 2001 Percentage removal of various constituents of Phase I and II wetlands

Pollutant	Phase II	Phase I	Phase I 2000 (Tilley et al., 2002)
NH_3-N	89 % ($p < 0.1$)	(227) %	(48 %)
TP	7 %	41 % ($p < 0.05$)	31 %
TSS	(26) %	30 % ($p < 0.1$)	65 %
VSS	(7)%	39 % ($p < 0.05$)	48 %
ISS	(49) %	27 %	76 %
BOD_5	(17) %	55 % ($p < 0.05$)	17 %

only Phase I was operating. The percentage removal for several constituents in the 2001 season is presented in Table 11.2.

Paired mean-difference testing was also performed on the inlet and outlet concentrations of each wetland. The deeper Phase II is effective at reducing NH_3-N (p value = 0.057). Conversely, the shallower Phase I was effective at removing total phosphorus (p value = 0.007), total suspended solids (p value = 0.020), volatile suspended solids (p value = 0.008) inorganic suspended solids (p value = 0.066), and CBOD (p value = 0.002). The NH_3-N increase in Phase I, while apparently large, (227 %) is an increase from only 0.0275 to 0.09 mg/l. A comparison with removals from Tilley et al. (2002) reveals a reduction in suspended solids removals but an increase in TP and BOD_5 removals for 2001.

Plotting mean concentration versus fractional distance illustrates the change in concentration within each type of wetland. Figure 11.6a and b depict ammonia-nitrogen profiles as a function of fractional distance for Phases II and I. Figures 11.7a, b and 11.8a, b show similar profiles for total phosphorus and 5-day BOD, respectively. The Y-error bars are plus or minus one standard deviation.

One-way ANOVA testing was performed for the five sampling locations in both wetlands. Pollutant reductions across Phase I are significant for total phosphorus, volatile suspended solids, and 5-day BOD ($p < 0.05$). Post hoc analysis using Scheffe's test indicate that the greatest removals were achieved using the whole wetland. Phase I demonstrated a 41 % removal efficiency for total phosphorus and a 39 % removal efficiency at the third interior sampling point indicating the last 20 % of the wetland only removed an extra 2 % of phosphorus (Fig. 11.7b). Five-day BOD had a 55 % removal for the whole of Phase I and a 54 % removal from the first interior sampling point to the outlet indicating that about the first 40 % of Phase I removed only 1 % of 5-day BOD (Fig. 11.8b).

The expanse of wetland from the inlet of Phase I to the first interior sampling point is essentially open wetland, and the area from the third interior sampling point to the outlet, is moderately vegetated while the interior of the wetland is heavily vegetated. Interestingly, for all three pollutants; total phosphorus, volatile suspended solids, and 5-day BOD, there were no significant removals from the inlet to the first interior sampling point, however, from the third interior sampling point to the outlet volatile suspended solids and 5-day BOD had greater than 10 % removals.

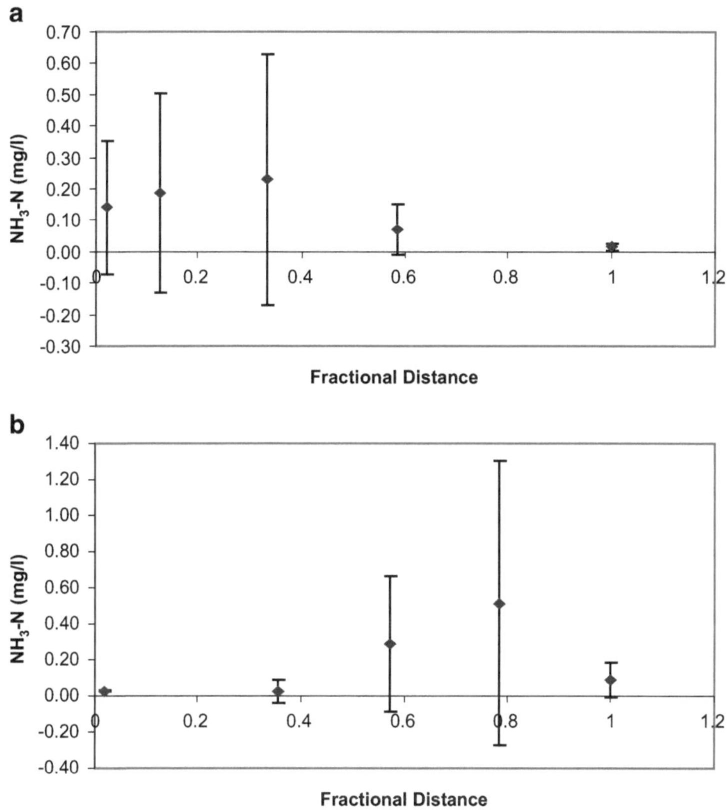

Fig. 11.6 Average 2001 (6 events) (**a**) Phase II wetland NH_3-N vs. fractional distance, (**b**) Phase I wetland NH_3-N vs. fractional distance

Based on these results the design of the new wetland following expansion (Phase III) was similar to a sequential wetland model referred to by other authors (U.S. Environmental Protection Agency [U.S. EPA], 1999, 2000). The recommended sequential model consisted of at least three alternating shallow and deep zones in the wetland. This design was developed to remove pollutants in the shallow zones, re-mix the effluent in the deep zone and remove more pollutants in the following shallow zone. This should reduce the effects of short-circuiting and allow for the sequential treatment of constituents similar to a wastewater treatment plant. The prior analysis indicated that the Phase I/II combination wetland behaved as two wetlands or as two zones in a sequential model. Phase II was deeper with submerged vegetation dominating. This wetland reduces dissolved constituents such as NH_3. Phase I demonstrated significant removals of parameters that were associated with particulate and settleable solids.

Since the existing wetland system appeared to perform as a sequential treatment system a proposal was made and approved to fully implement the eco-technology at

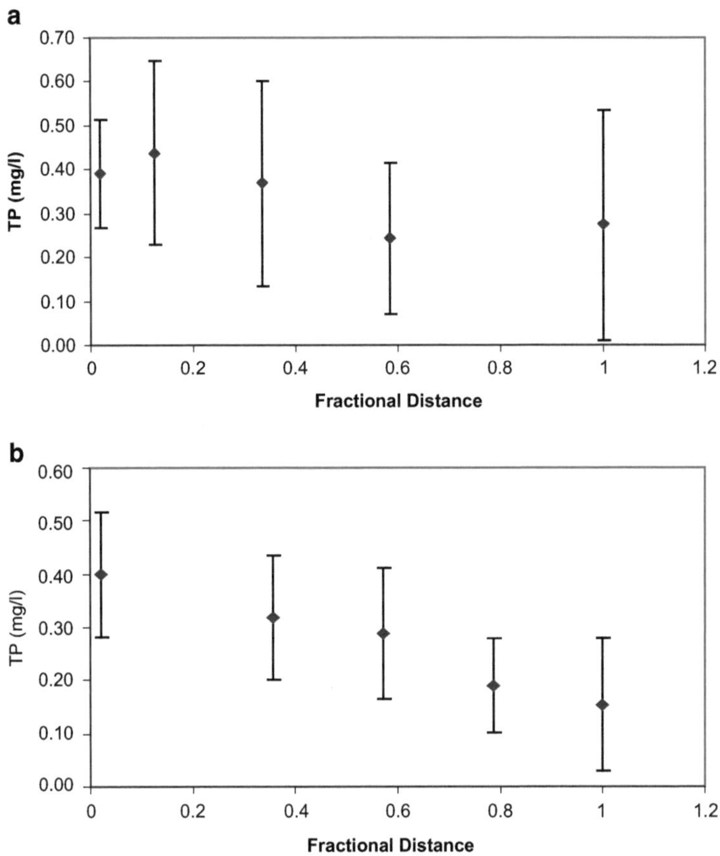

Fig. 11.7 Average 2001 (6 events) (**a**) Phase II wetland total phosphorus vs. fractional distance, (**b**) Phase I wetland total phosphorus vs. fractional distance

LASAF during the planned 2003 expansion project. The new Phase III constructed wetland was constructed in 2003 and operational for the 2003 and 2004 growing seasons.

Table 11.3 lists the results of analysis for the removal of nutrients in the Phase I, II and III Loma Alta wetlands systems before and after the inundation of 2002. The table displays the average inlet concentration and removal percentage for carbonaceous biochemical oxygen demand (CBOD), total suspended solids (TSS), inorganic suspended solids (ISS), soluble reactive phosphorous (SRP), total phosphorous (TP) and ammonia (NH3) over the years 2001 through 2004. Two comparisons are possible, that of the treatment by the Phase I and II total wetland treatment system (19 ha) before and after the inundations of 2002, and that of the Phase I and II system with the smaller sequential Phase III (8 ha).

The results show that the flooding of the wetland system not only affected the vegetation distribution (as listed in Table 11.1) but also diminished the treatment

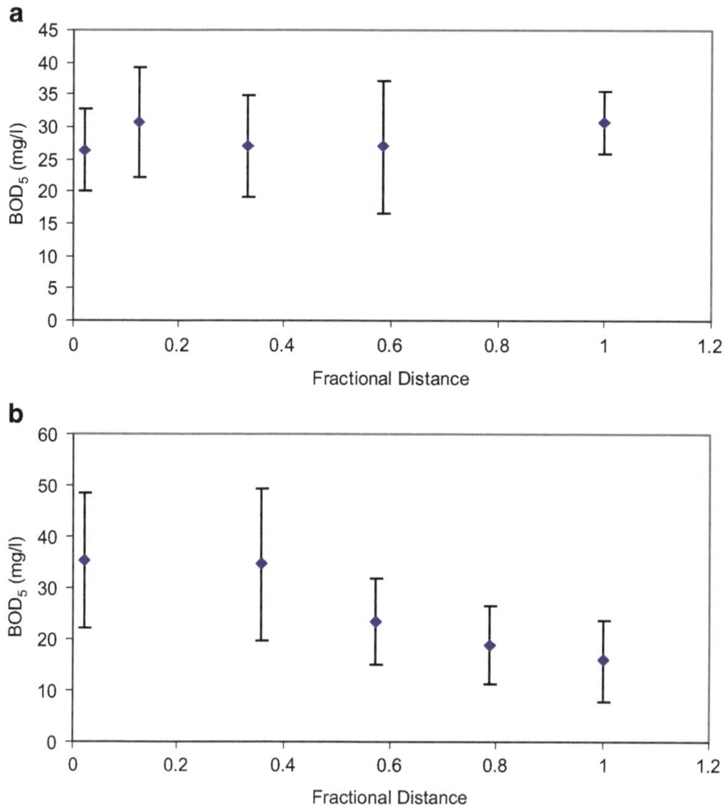

Fig. 11.8 Average 2001 (6 events) (**a**) Phase II wetland BOD_5 vs. fractional distance (**b**) Phase I wetland BOD_5 vs. fractional distance

efficiency in 2002. Increased scouring in connecting channels and canals caused by more pumps brought on line to move the larger volume of water through the system may have caused some of the increase in TSS. Water turbidity was also much higher during this season. However, in 2003 the wetlands had begun to recover and treatment efficiency for almost all of the parameters had been restored, even though the Phase I and II wetlands were now dominated by forms of more submerged vegetation, primarily widgeon grass (*Ruppia maritima*). By the 2004 season, the steady state recirculating water quality had improved to 2001 levels and removal percentages were sufficient to meet the operator's water quality goals. Since 2005, the shrimp aquaculture operation has been suspended at LASF in anticipation of the need for higher commodity prices and perceived shift in demand for other farmed species such as redfish.

Change-point analysis was proven to be useful as a tool to evaluate the onset of more dynamic flow conditions during harvesting periods for the aquaculture operations. For the 2003 and 2004 years of operation the smaller Phase III

Table 11.3 Summary parameter analysis for quasi-steady state aquaculture wetland operation over the seasons 2001–2004

			Phase I and II wetlands				Phase III wetlands		
Season	Parameter	# of sampling events	Mean inlet (mg/L)	Mean % removal	% Standard deviation	# of sampling events	Mean inlet (mg/L)	Mean % removal	% Standard deviation
2001	CBOD	6	30	54	32	–	–	–	–
	TSS	6	102	30		–	–	–	–
	ISS	6	61	61		–	–	–	–
	SRP	6	0.18	37		–	–	–	–
	TP	6	0.34	55		–	–	–	–
	NH3	6	0.14	16		–	–	–	–
2002	CBOD	4	30	7	20	–	–	–	–
	TSS	4	111	−47		–	–	–	–
	ISS	4	79	14		–	–	–	–
	SRP	4	0.04	17		–	–	–	–
	TP	4	0.64	33		–	–	–	–
	NH3	4	0.33	65		–	–	–	–
2003	CBOD	10	28	24	38	11	29	5	5
	TSS	10	142	44		11	110	25	
	ISS	10	77	61		11	71	55	
	SRP	10	0.11	59		11	0.09	30	
	TP	10	0.68	20		11	0.67	15	
	NH3	10	0.11	28		11	0.11	15	
2004	CBOD	11	23	45	43	13	23	23	36
	TSS	11	131	28		13	151	42	
	ISS	11	69	40		13	91	62	
	SRP	11	0.16	59		13	0.14	33	
	TP	11	0.44	19		13	0.39	43	
	NH3	11	0.17	24		13	0.22	40	

sequential wetland system was able to closely match the efficiency of the larger Phase I and II system with an average removal of 39 % of TSS removal for Phase III compared to 43 % for Phases I and II, 60 % of ISS for Phase III compared to 59 % for I and II, 52 % for SRP for Phase III compared to 45 % for I and II, 60 % removal of TP for Phase III and 56 % for I and II, 44 % ammonia removal compared to 60 % for Phases I and II, and 49 % CBOD removal for Phase III with 60 % for Phases I and II. The CBOD removal in Phase III has shown steady improvement as more emergent vegetation has filled in. CBOD removal in these aquaculture systems may be problematic in some instances as much of this parameter is apparently associated with healthy algal species, as evidenced by a strong correlation between chlorophyll-a and CBOD in wetland water samples. It would be expected for such natural waters to maintain a threshold CBOD value.

11.4 Conclusions

Several conclusions were reached for this aquaculture wetland treatment system performance and recovery after the flood inundation of 2002.

- Despite variable loadings from pond discharges into aquaculture wetlands, a steady state operation developed prior to harvest operations can be used to evaluate the nutrient removal performance of the wetlands early in the season. New models to deal with the unique dynamic loadings in harvest operations aquaculture wetlands and storm water wetlands will be needed to better evaluate wetland performance during these conditions.
- It is suggested here that techniques such as change-point analysis can help lend insight as to when the retention time significantly changes in a dynamic wetland system as a result of changing water depth and provide clues as to the nature of its hydraulic efficiency.
- The flooding events caused by storms and operator management requirements in 2002 resulted in diminished vegetation and wetland performance which was able to be restored in 2003, with the growth of primarily submerged vegetation.
- The smaller (8 ha) Phase III sequential basin design appeared to be able to effectively treat aquaculture effluent as well as the larger (19 ha) dual system Phase I and II wetlands.
- Close attention to water depth and recirculation rate management suggests a linkage to optimal performance for water treatment targets. Holland et al. (2004) also found that increasing wetland water depth caused greater short-circuiting and direct changes to the residence time distribution.

Acknowledgements This research was supported in part by the National Science Foundation sponsored CREST program at Texas A&M University-Kingsville for Research on Environmental Sustainability for Semi-Arid Coastal Areas (RESSACA) number HRD-0206259, and by Loma Alta Aquaculture LTD. Any opinions, findings, and conclusions or recommendations expressed in this material are those of the author and do not necessarily reflect the views of the National Science Foundation or Loma Alta LTD.

We would also like to acknowledge the contributions of Ron Rosati from the Texas A&M University-Kingsville College of Agriculture, David Tilley at the University of Maryland, Jerry Taylor of El Sauz Ranch, Robert Smiley, Keith Gregg and Fritz Jaenike of Harlingen Shrimp Farms. The researchers would also like to express their gratitude to the H. P. Pinnell family, the El Sauz Ranch, and Loma Alta Aquaculture LTD, for permitting the research to be conducted at the aquaculture facility wetland.

References

APHA. (1992). *Standard methods for the examination of water and wastewater* (18th ed.). Washington, DC: American Public Health Association, American Water Works Association, Water Environment Federation.

Carleton, J. N. (2002). Damkohler number distributions and constituent removal in treatment wetlands. *Journal of Ecological Engineering, 19*, 233–248.

Costa-Pierce, B. A. (1998). Preliminary investigation of an integrated aquaculture-wetland ecosystem using tertiary-treated municipal wastewater in Los Angeles County, California. *Ecological Engineering, 13*, 341–354.

Coveney, M. F., Stites, D. L., Lowe, E. F., Battoe, L. E., & Conrow, R. (2002). Nutrient removal from eutrophic lake water by wetland filtration. *Journal of Ecological Engineering, 19*, 141–159.

Hach. (1991). *HACH/DR2000 spectrophotometer procedures manual*. Lovewland, CO: HACH Company.

Hinkley, D. V. (1971). Inference about the change-point from cumulative sum tests. *Biometrika, 58*, 509–523.

Hinkley, D., & Schechtman, E. (1987). Conditional bootstrap methods in the mean-shift model. *Biometrika, 74*, 85–93.

Holland, J., Martin, J., Granata, T., Bouchard, V., Quigley, M., & Brown, L. (2004). Effects of wetland depth and flow on residence time distribution characteristics. *Journal of Ecological Engineering, 23*, 189–203.

Jensen, G. (2005). *Update on aquaculture trends and select USDA aquaculture programs*. Paper presented at the 35th annual meeting of the Texas Aquaculture Association, Corpus Christi, TX.

Kadlec, R. H. (2000). The inadequacy of first order treatment wetland models. *Journal of Ecological Engineering, 15*, 105–119.

Kadlec, R. H., & Knight, R. L. (1996). *Treatment wetlands*. Boca Raton, FL: Lewis Publishers.

Lin, Y. F., Jing, S. R., Lee, D. Y., & Wang, T. W. (2002). Nutrient removal from aquaculture wastewater using a constructed wetlands system. *Aquaculture, 209*, 169–184.

Manly, B. F. J. (2001). *Statistics for environmental science and management*. Boca Raton, FL: CRC Press.

Moss, S. M. (2002). Marine shrimp farming in the western hemisphere: Past problems, present solutions, and future visions. *Reviews in Fisheries Science, 10*, 601–620.

Paez-Osuna, F. (2001). The environmental impact of shrimp aquaculture: A global perspective. *Environmental Pollution, 112*, 229–231.

Resampling Stats. (2002). *Microsoft Excel Add-in* (Version 2.0). www.resample.com

Summerfelt, S. T., Adler, P. R., Glenn, D. M., & Kretschmann, R. N. (1999). Aquaculture sludge removal and stabilization within created wetlands. *Aquacultural Engineering, 19*, 81–92.

Taylor, W. (2000). *Change-point analyzer 20 shareware program*. Libertyville, IL: Taylor Enterprises. http://www.variation.com/cpa.

Tilley, D. R., Badrinarayanan, H., Rosati, R., & Son, J. (2002). Constructed wetlands as re-circulation filters in large-scale shrimp aquaculture. *Aquacultural Engineering, 26*, 81–109.

Treece, G. (2005). Texas Aquaculture Association. http://www.texasaquaculture.org.

U.S. Environmental Protection Agency [U.S. EPA]. (1999). *Free water surface wetlands for wastewater treatment technology: A technology assessment* (EPA Rep. 832-S-99-002). Washington, DC: Office of Water.

U.S. EPA. (2000). Constructed wetlands treatment of municipal wastewaters. In *The EPA manual* (EPA/628/R-99/010). Cincinnati, OH: Office of Research and Development.

Index

D. Ramirez et al. (eds.), *Environmental Sustainability Issues in the South Texas–Mexico Border Region*, DOI 10.1007/978-94-007-7122-2,

MIX
Papier aus verantwortungsvollen Quellen
Paper from responsible sources
FSC® C105338
FSC
www.fsc.org

If you have any concerns about our products,
you can contact us on
ProductSafety@springernature.com

In case Publisher is established outside the EU,
the EU authorized representative is:
Springer Nature Customer Service Center GmbH
Europaplatz 3, 69115 Heidelberg, Germany

Printed by Libri Plureos GmbH
in Hamburg, Germany